Salmon Production, Management, and Allocation

Salmon Production, Management, and Allocation
Biological, Economic, and Policy Issues

Edited by
William J. McNeil

Oregon State University Press
Corvallis, Oregon

Publication of these proceedings has been supported by

NCRI
National Coastal Resources Research and Development Institute

The goal of NCRI is economic diversification and development of coastal communities through compatible multiple uses of marine and coastal resources.

Library of Congress Cataloging-in-Publication Data

Salmon production, management, and allocation.
    Based on papers presented at the World Salmonid Conference held Oct. 2-3, 1986, in Portland, Or. and sponsored by the Salmonid Foundation.
    Includes bibliographies and index.
    1. Salmon—fisheries—Congresses.   2. Salmon—Congresses.   I. McNeil, William J.   II. World Salmonid Conference (1986 : Portland, Or.)   III. Salmond Foundation.
SH346.S25   639.3'755   87-22135
ISBN 0-87071-354-X (alk. paper)

Preface

The World Salmonid Conference was organized to facilitate communication among professional disciplines to provide essential guidance for the development of salmonid sport and commercial fisheries and aquaculture. These proceedings provide permanent documentation of the conference for reference by policymakers, regulatory agencies, legal experts, scientists, scholars, students, and the public.

The Salmonid Foundation was pleased to provide funding for expenses of invited participants and other costs associated with the conference. Selection of participants was the responsibility of the convener, Dr. William J. McNeil, Professor of Fisheries, Oregon State University.

The Salmonid Foundation promotes the sporting aspects of angling while simultaneously encouraging conservation and development of salmonid stocks and the protection of their habitat. We are concerned that society's ability to resolve conflicts over competition for scarce resources and to maintain and expand the long-term productivity of our salmonid fisheries has been impaired by a maze of regulations resulting from the large number of international, federal, state, tribal, and local agencies with authority over salmonid resources and/or their habitat. We believe that policies must be established to ensure that maximum values accrue to society from our salmonid resources. Our main purpose in supporting this conference was to facilitate the acquisition, evaluation, and dissemination of information to form the basis for such policies. We are confident that the interests of recreational fishers and other users of salmonids will be best served by informed evaluation of the biological and economic benefits of policies designed to conserve and allocate scarce resources and to expand the resource base through aquaculture.

Chuck Voss
Executive Director
Salmonid Foundation

Acknowledgments

Cost of the World Salmonid Conference, including publication of these proceedings, was shared by a number of groups and individuals. Arrangements to hold the conference, including travel for invited participants, were made by the Salmonid Foundation. Expenses directly associated with the meeting were paid by the foundation, which received contributions from the American Fishing Tackle Manufacturers Association, E. Hadley Stuart, Jr., Larry Schoenborn, and Lamiglas, Inc. A gift from Tom Becker to the Oregon State University Foundation was used to pay for illustrations. Other publication costs were subsidized by the National Coastal Resources Research and Development Institute.

Jo Alexander, Managing Editor of the Oregon State University Press, organized and managed publication of these proceedings. She dedicated many hours of her outstanding editorial skills to the project. She deserves to be listed as coeditor, but she insists that she was only doing what was expected of her. However, I can state without qualification that quality of this publication has been significantly enhanced by her efforts. Any expression of gratitude on my part is inadequate recognition of her contributions.

William J. McNeil

Contents

Foreword

Salmonid fishes occur naturally or have been transplanted into all major temperate marine and freshwater basins of the world. They rank very high among the most valued food fishes.

World harvest of salmon began to increase in the early 1970s after having declined to less than one-half of historic levels. Recovery has been rapid, and historic high levels are once again being approached. If the trend of increasing supply continues, salmon will soon be more plentiful than ever before.

How can the rapid turnaround in the market supply of salmon be explained? Contributing factors appear to include:

—Reduced high seas fishing on mixed stocks.
—Increased spawner escapement.
—Improved environmental conditions in Alaska.
—Increased production from aquaculture.

High seas fisheries expanded rapidly in the north Pacific and north Atlantic oceans in the 1950s and 1960s. These fisheries exploited mixed stocks and imposed additional fishing mortality on stocks that were already being harvested by coastal fisheries. The resulting depletion of salmon stocks was widespread throughout the north Pacific and north Atlantic. International agreements to reduce harvest of salmon on the high seas began to take effect in the late 1960s, but substantial reductions were not apparent until the 1970s.

Long-term research to determine escapements of natural spawners required to maintain high population densities of salmon began to receive emphasis in the 1950s. Information that is essential for establishment of escapement goals continues to be gathered, but fishery managers now possess sufficient knowledge to define realistic escapement goals for many important stocks. Regulation of fishing effort to insure adequate escapements has contributed to recovery of natural production in many areas where freshwater habitat has not been seriously degraded.

Alaska stocks have flourished in recent years, and record catches have been common. Scientific evidence is accumulating that freshwater and marine environmental conditions improved for Alaska stocks during the last decade. Because Alaska is the major world producer of salmon, the positive impact on supplies has been substantial.

Even though natural production may have increased in recent years, aquaculture is probably the major contributor to the growing supply of salmon. Salmon aquaculture has experienced unprecedented growth in recent decades. Industrial-scale ranching[1] and farming[2] are making substantial progress in both northern and southern hemispheres. Based on numbers of ranched salmon released from hatcheries and the volume of farmed salmon entering world markets, I am of the opinion that nearly 40 percent of today's world harvest originates from aquaculture (see *Afterword*, page 187). Projection of recent trends suggests that aquaculture could provide more salmon than natural production by the end of this century.

Salmon aquaculture is practiced on an industrial scale in three oceanic basins and their contiguous seas—north and south Pacific and north Atlantic. There is also limited potential for salmon aquaculture in the south Atlantic Ocean (Argentina and the Falkland Islands) and the Indian Ocean (Tasmania).

Industrial salmon ranching continues to expand rapidly on both sides of the north Pacific Ocean. Ranching is much less extensive in the north Atlantic and south Pacific oceans than in the north Pacific, but is growing steadily. Nearly five billion smolts are currently released from hatcheries into marine and freshwater nursery waters throughout the world. This number has been doubling about every 10 years, and the immediate outlook is for continued expansion at about the same rate.

---

[1] A salmon ranch releases smolts from hatcheries into natural waters.
[2] A salmon farm grows salmon to market size in captivity.

The salmon farming industry has become well established in the western north Atlantic Ocean, and major new initiatives are emerging in the north and south Pacific Ocean. World production of farmed salmon is currently about 50,000 mt and has been doubling in 3 years.

Ranching presently contributes a much larger biomass of salmon than farming, but the value of farmed salmon relative to ranched salmon is much greater than production levels indicate. This is because the principal ranched species (pink and chum salmon) have a low value per unit of weight in comparison to the principal farmed species (Atlantic, coho, and chinook salmon). Pink and chum salmon are released at a much smaller size than other species, and they are inexpensive to produce in hatcheries. Furthermore, their ocean survival rates compare favorably with species that enter marine waters at a much larger size. These are the primary reasons why there is an economic advantage to ranch them and why pink and chum salmon constitute 90 percent or more of juvenile salmon released for ranching.

The advent of industrial-scale salmon aquaculture manifests a transition from hunting to aquaculture in some economies. Such a transition is well advanced in Japan, which has become the world's leading producer of ranched salmon and a major producer of farmed salmon. For many years Japanese fishers were primarily hunters of naturally-produced salmon on the high seas. Because mixed stocks from several countries were targeted for hunting, international opposition forced Japan to greatly reduce high seas catches. Japan compensated by expanding salmon aquaculture. Today, the biomass of salmon harvested in Japan from aquaculture exceeds their harvest of salmon on the high seas and has become the primary source of supply for Japanese markets.

Recreational fisheries for salmon are more important contributors to economic activity than commercial fisheries in some geographic areas. Allocation of salmon between sport and commercial fishers is an important policy issue which relates at least indirectly to aquaculture. Iceland and New Zealand, for example, have encouraged commercial aquaculture to provide for production of market fish, but reserve the hunting fisheries for recreational uses.

Rapid growth of salmon aquaculture and its biological, economic, and policy implications to production, management, and allocation of salmon stimulated interest in a conference involving experts from several disciplines. The resulting "World Salmonid Conference" was sponsored by the Salmonid Foundation in cooperation with Oregon State University. Conferees convened October 2 and 3, 1986, in Portland, Oregon. Twenty papers covering a broad range of pertinent topics are included in these proceedings.

The World Salmonid Conference provided a forum for fisheries scientists and managers, aquaculturists, economists, legal experts, and public policy analysts to address issues of importance to the integration of aquaculture in the overall production of salmon. The multidisciplinary group of conferees was asked to address questions ranging from salmon ecology to social policy. Some authors emphasize interpretation of ecological data, while others assess biological, economic, and policy issues. Results of scientific investigations on migrations, distribution, and marine survival of salmon are included because such research provides a foundation for conserving and enhancing stocks of salmonids and for understanding the capacity of marine waters to grow salmon. Rapid growth of salmon aquaculture has provided a stimulus for assessments of mixed-stock fisheries and management of disease and genetic make-up of stocks. These important topics are considered by several authors. Complex economic and social issues are also addressed from several perspectives, including technology innovation, economics, law, and resource allocation among user groups.

The keynote address by W. F. Royce outlines goals for the conference and discusses diversity in sources of production and management of salmon resources of the northeastern Pacific Ocean. World trends in producing and utilizing salmon are described by J. E. Lannan, who also discusses the compexity of institutional policies. A second paper by W. F. Royce provides a chronology of salmon production in the northeastern Pacific and offers hypotheses about factors contributing to the recent recovery of salmon production in Alaska.

Release of ranched salmon into public waters raises challenging legal questions about ownership and rights of fish. A. A. Hampson assesses the legal framework for ranching (from the perspective of a salmon) under common and legislated law.

Two papers give examples of industrial-scale salmon aquaculture. Japanese ranching and farming industries are assessed by Y. Nasaka and the Norwegian farming industry by F. Gjerset. Japan, the world's largest consumer of salmon, now satisfies about one-half of its domestic demand through ranching. More farmed salmon are grown in Norway than all other countries combined.

Distribution and survival of salmon in marine waters is discussed by several authors. Ocean distributions of salmon stocks in the north Pacific and impacts of high seas fisheries on stocks are analyzed by C. K. Harris, who also describes international agreements to manage high seas fisheries. W. G. Pearcy summarizes data from ocean sampling of coho salmon off Oregon and Washington and discusses hypotheses about variable marine survival.

Much has been said about good hatchery practice and high quality smolts. R. Gowan reports, however, that manipulations (e.g. size, time, and location of release) designed to meet conditions external to the hatchery may be more important for marine survival than practices for growing smolts within the hatchery.

Observed interactions between pink and chum salmon and hypothesized implications about the capacity of marine waters to grow these species are presented by E. O. Salo. Observations on marine migrations and survival of salmon from Bristol Bay, Alaska, the world's largest natural producer of salmon, are analyzed by D. E. Rogers.

Conservation issues related to harvest of salmon by fisheries operating on mixed stocks are addressed by two authors. Harvest rates on stocks which intermingle on fishing grounds need to achieve a balance between conservation and social goals, according to D. E. Bevan, who favors a flexible approach to harvest management utilizing daily catch reports to monitor fisheries. Among several important interactions between hatchery and natural stocks discussed by C. J. Walters is the problem of managing fisheries in mixed stocks of hatchery and naturally produced fish. Because hatchery stocks typically have a lower escapement requirement (i.e. a higher egg-to-adult survival) than natural stocks, there is a tendency to overfish natural stocks.

Allocation of salmon among user groups is addressed in two reports. The importance of allocation decisions for realizing maximum economic benefits from salmon introduced to the Great Lakes is emphasized by H. A. Tanner, who describes biological and economic benefits from introduc-

tions of salmon. Roles of "conservation" and "allocation" in the fishery management decision process are differentiated by C. L. Smith.

Economic issues are primary subjects of three reports. Economic impacts of commercial and sport caught salmon are evaluated for a U.S. west coast fishery by C. N. Carter and H. D. Radtke to assist decisions on allocation. Conditions influencing markets for salmon are described by R. S. Johnston, who contrasts trends in the total economy (macroeconomics) with trends in the salmon supply sector (microeconomics). The overall impact of sport fishing on the U.S. economy is assessed by D. B. Rockland.

Aquaculture operations invariably involve transplantation of stocks and associated risk of spreading infectious diseases. This important issue is reviewed by J. S. Rohovec, J. R. Winton, and J. L. Fryer, who make recommendations about practical means to minimize risk.

Breeding strategies affect the genetic make-up of aquaculture stocks. Strategies appropriate for ranching and farming are differentiated by W. K. Hershberger.

William J. McNeil
Professor of Fisheries
Oregon State University

Our Outdated Policies

William F. Royce

Professor Emeritus
School of Fisheries
University of Washington

The salmonidae have been called the most valuable family of fishes in the world. They are also the most fascinating to many people because of their anadromous habit. Anthony Netboy and Roderick Haig-Brown, in their marvelous books and articles about the Pacific and Atlantic salmons, caught our sense of wonder about their life that starts in stream gravel and includes incredible ocean migrations, as well as our deeply-rooted sense of caring for them and using them sensibly.

As we review the role of salmon in the great transition of our market fisheries from fishing to fish farming, we will remember our history of cherishing these remarkable fish. To many, they are symbolic of an obligation to our environment. We will discuss the anadromous salmons of the Pacific, the most abundant and most valuable members of the family, and we also will draw on our long experience with other important salmonids, the chars and trouts. Some of these live entirely in fresh water but share similar life habits with the Pacific salmons. We will be concerned with the future supply of salmon for the market, but we will also find ourselves involved with use of the salmon for recreation, or for subsistence, or just for watching by people who marvel at their natural life.

These concerns will entangle us in three kinds of issues: biologic, economic, and policy. Each of these is customarily dealt with in different arenas, but the papers in this collection are concerned with all three. Issues related to economics and policy are customarily dealt with in the legislature and the marketplace, but they also deserve a more thorough scholarly examination.

Some sovereignties, including Canada and the United States, have a strong policy commitment to conservation supported by scientific evidence. But policy is constantly modified by needs of users, both commercial and recreational, who politically influence state, provincial, and federal legislators. Unfortunately, the overall marine fishery policies of both countries since the middle 1970s have been shaped by expectations of major benefits from the extension of jurisdiction over fisheries resources to 200 miles, which occurred in 1977. Both have had commercial fishery development programs that have produced few if any benefits. I believe that the failure lies at least partly in a lack of public realization that the world's fish businesses are rapidly changing. National and state or provincial policies have tended to react to old perceptions of needs that may no longer be pertinent. To continue as we have in the past is a policy of defeat. We must expect no final solutions—only hope to develop the best evolutionary process in our democratic societies.

Let me discuss briefly some fundamental facts and policy issues.

OUR DIVERSE SALMON FISHERIES

Market fishing in the Pacific Northwest takes all five species, but the great majority are the invertebrate feeders, pink, chum, and sockeye.[1] Recreational fishers share the fish feeders, coho and chinook (and very minor catches of the others) with market fishers and, in addition, have nearly exclusive use of the fish feeding trouts and chars.

Market catches of all species of Pacific salmon in North America have recently averaged more than 340,000 mt annually. Total recreational and subsistence catches are much less—recreational catches probably total only about 2 percent, subsistence catches only 3-4 percent of the market

---

[1] See Appendix on page 189 for a listing of the scientific and common names of fish mentioned in this volume.

catches—but these small fractions are much more important, geographically and politically, along the coast from Southeast Alaska to California.

A DIFFERENCE IN MANAGEMENT

Management of commercial net fishing for salmon in North America, mostly without reliance on hatcheries, recently has become spectacularly successful. The recent catch total has reached a record about 50 percent larger than catches during the 1950s, 1960s, and 1970s, when production averaged less than 225,000 mt annually.

This restoration has been a triumph of good research and management of net fishing, principally in British Columbia and Alaska. It is based on the fundamental principle of restricting net fishing to waters inside the "surf line," where the stocks begin to segregate. Catch is regulated for individual stocks to the maximum extent compatible with maintenance of good flesh quality prior to maturation.

The management of high seas trolling for market, subsistence, or recreational fishing does not allow nearly as good control of catch and escapement. The stocks are mixed; they migrate through multiple jurisdictions; their regulation is largely the result of confrontational politics; and they are subject to considerable waste due to capture of immature fish plus loss of injured fish. The difficulties continue despite major, costly programs of research, management, and enhancement. These fisheries, which are especially important for recreation because they involve favorite species and are close to population centers, are a major concern.

OUR SALMON FISHING CONSTITUENCIES

Salmon policies in Canada and the United States are strongly influenced by recreational fishers, mainly because recreational fishers as a group outnumber all market fishers by about 300 to 1. They are concerned, of course, primarily with the recreational species, and they also strongly support preservation of the environment.

Recreational salmon fishers in the Pacific Northwest from Alaska to California probably number more than one million, whereas subsistence fishers probably number less than fifty thousand, and commercial salmon trollers less than twenty thousand. Even so, a majority of the licensed commercial trollers appear to be hobby fishers, rather than fishers who make most of their living from fishing salmon plus other species. Those who expect fishing to provide the major part of their livelihoods probably number less than one thousand.

THE VALUES OF SALMON

The total monetary value of salmon as landed for the market in the United States (recently $350-$400 million annually) is second only to the value of shrimp. It is about one-third of the total value of finfish, and about one-sixth of the total value of all fish, including shellfish. Only about 10 percent of this market value is provided by the commercial troll catches, the balance by the salmon net fisheries, most of which land in Alaska. The total monetary value of salmon products is probably about double the value of the catch as landed.

Comparison of the value of market catches and recreational fishing must be made on a more subjective basis such as estimates of the impact on local personal income. Such estimates for salmon fishing in the ocean off California to Washington in 1985 show that ocean trolling for the market enhanced local personal income about $47 million, salmon angling about $28 million. The value of salmonids caught by anglers also has been estimated variously according to amounts spent per fishing day or per fish caught, and lost opportunities for fishing or for salmon to spawn. Such estimates have valued salmon up to more than one hundred dollars each for capture, and up to more than five hundred dollars each for spawners escaping up the Columbia River. The value of salmonids caught by Indians might be regarded as comparable to market value if they are sold, but Indian representatives have asserted that the cultural value cannot be expressed in monetary terms. Such differences in value defy blending by any analytical system and lead to allocation based largely on politics.

RANCHING AND FARMING
Public ranching

Pacific salmon ranching with government supported hatcheries began more than a century ago in North America. Systems were adopted and abandoned in Alaska and British Columbia but continued in the states of Washington, Oregon, and California. Recently ranching has been resumed in British Columbia, and also in Alaska but with a state provision that some hatcheries are to be operated by nonprofit corporations with about half their financing provided by assessments on the market fishers who benefit.

Almost all of the salmonid hatchery operations from British Columbia south to California (except those for freshwater trout) support the fisheries for steelhead, chinook, and coho. They now sustain about half the fisheries for those species outside Alaska. The nonprofit hatcheries in Alaska, however, primarily support pink and chum salmon fisheries. The advanced fry of these species can be released to go to sea with little or no feeding—a major item of cost in the culture of chinook, coho, or steelhead. The annual catches attributed to these pink and chum hatcheries have increased rapidly since the late 1970s and have recently exceeded 15,000 mt (about eight million fish).

Even more spectacular results have been obtained by the chum salmon ranching industry in Japan (see paper by Nasaka, this volume). Coastal chum catches, once averaging more than 20,000 mt annually, had declined to about 10,000 annually in the 1960s when the ranching system was initiated. Coastal chum runs have since expanded to yield more than 120,000 mt annually in the mid-1980s.

Private ranching

Several businesses in the United States have attempted private ranching but all have been discouraged by the complexity of the problems and by political opposition. It is unquestionably a risky business, especially if stocks of salmon have to be moved from distant streams to the hatchery. Then the hatchery fish have to be bred for a few generations in order to develop the ability to find ocean feeding grounds and return. If political roadblocks are thrown up in addition, as they have been in the United States, prospects appear to be doomed.

Private farming

Estuarine pen farming of Atlantic salmon, coho, and rainbow trout is burgeoning in at least a dozen countries. It is attracting investment wherever there are clean sheltered waters with good temperatures the year around. The technology and favorable economics have been proven.

Norway has been the leader, with production expected to reach about 50,000 mt in 1986, but all of the northwest European countries plus Japan, Chile, New Zealand, Canada, and Australia expect rapidly increasing production. According to recent informed reports, about one hundred million dollars of private investment has already been committed to pen culture in British Columbia.

SALMON MANAGEMENT
Public Costs

Management in the business sense involves a recurring comparison of expenditures and income with a continuing projection of expectations. Government fishery management systems do not operate in this way. A major task of Fishery Management Councils on the U.S. west coast is to coordinate the management of salmon fisheries. The Councils publish detailed statistics on catches and participation, but they are silent both on comparisons of public costs and benefits and on evaluation of long-term trends in recreational and business aspects of the fisheries.

When I studied a 1985 review of the ocean salmon fisheries prepared by the Pacific Fishery Management Council, I noticed that about 10 percent of the commercial troll fishers made about half of the catches, about 40 percent of the fishers made about 90 percent of the catches, and some with commercial licenses made no commercial catches. I also recalled that an acquaintance of mine, a wholesale furniture salesman, had a salmon trolling license a few years ago, which he apparently used so that he and his son could have fun and claim tax losses.

I asked several people familiar with Council activities an apparently simple question: How do salmon trollers make a living when they can fish for salmon only a month or so each year? No one knew, but some guessed that a few were professional fishers who fished for albacore, crab, or other species at different times. I suspect that the others, comprising a large proportion of the commercial salmon trollers, should be designated as hobby fishers.

Management of this fishery and the associated recreational and Indian fisheries is expensive. No complete data are available, but estimates of partial costs for salmon and steelhead management, enhancement, enforcement, and administration by state, federal, and Indian agencies for fiscal years 1981 to 1984 were reported to the Salmon and Steelhead Advisory Commission as being in the region of $85 million annually. The total has been estimated to be perhaps 50 percent greater by people familiar with the system. These costs applied to all species of salmon and the steelhead, but presumably the management of net fisheries in inside waters for pink, chum, and sockeye required less than 10 percent of the total. The recent recreational and commercial troll catch of coho, chinook, and steelhead south of the Canadian border has been on the order of four million fish annually, giving a cost per fish on the order of at least twenty dollars. A comparison with Alaskan salmon management—of both recreational and commercial fisheries—is appropriate. According to informal but reliable reports, the total cost of salmon management there is about $35 million annually. With about 150 million fish in the catch (about 99 percent in the inside net fisheries) this is about 25¢ per fish.

The efficiency of management

Technical efficiency needs to be judged as well as costs. Alaskan salmon net fishing is a model, both technically and economically. On the other hand, management personnel from the fishery agencies recognize that management of ocean fishing on mixed stocks is grossly inefficient. Those fisheries operate on mixed stocks which can be identified in the catches only roughly by continuing and expensive tagging programs. Catches in mixed stock ocean fisheries also include a large proportion of immature fish, and substantial numbers of sublegal fish which are hooked and released fail to survive.

Perpetuation of commercial ocean fisheries on mixed stocks of salmon off California, Oregon, and Washington and implementation of a new salmon treaty with Canada in order to better control the catches from the mixed stocks is an extension of an existing wasteful system in which the real costs to the public are concealed. I am aware of no breakthroughs in research that indicate the likelihood of any major improvement in conservation of stocks exposed to these fisheries.

TRENDS IN THE INTERNATIONAL FISH TRADE

The salmon business has long been an important part of international fish trade. Two trends are important to us.

According to reports from the Food and Agriculture Organization (FAO) of the United Nations, the volume of the total international trade in fish products remained about the same between the early 1970s and the early 1980s, but the value quadrupled from about $4,000 million to $16,000 million annually (which was about twice the rate of inflation). Meanwhile, despite the U.S. fishery development program starting in 1979, the United States remained a major importer. U.S. imports attained new records annually to reach $5,883 million in 1984. By comparison, total exports in 1984 were valued at only $949 million—of which more than half were salmon products.

The increase in value of internationally traded fish products reflects the extraordinary abilities of large transnational companies to plan, develop products and markets, and finance operations. For example, Taiyo Fisheries, one of the three fish companies in Japan currently on the Fortune 500 list of the largest industrial corporations outside the United States, had annual sales for fiscal year ending January 31, 1985 of $4,700 million—twice the total value of the entire U.S. catch as landed in 1984.

THE OVERALL ISSUE

Development of salmon farming or ranching in North America is just one of the opportunities for fisheries development which both Canada and the United States are missing, even though they have close to 15 percent of the total marine fish resources of the world.

I cannot summarize the problems any better or more pungently than Bart Eaton, a commercial fisherman of Issaquah, Washington, did at the Anchorage conference on fisheries management in the fall of 1984. He started with a quotation from the economist, John Maynard Keynes: "The difficulty lies not in new ideas, but in escaping from the old ones." He continued with a general comment on U.S. experience under the Fisheries Conservation and Management Act.

These are examples of a politicized management system, one in which expediency often undermines both business and biology. The current system, rife with inefficiencies and confusion, wastes the full potential of an industry that could provide far greater benefits to society . . . The first step is to establish a guiding philosophy so that our management policies, now largely backward-looking and intended to redress past inequities, point to a stable future. This managerial creed could provide the consistency now lacking in much of what our management system undertakes, and consistency is crucial to business.

Our task, and that of others to follow, might well be described as escaping from the old ideas and planning for the future of the salmon and the salmon businesses that shape our use of them.

Contemporary Trends in World Salmon Production and Management Policy

James E. Lannan

Professor of Fisheries
Oregon State University

Many in salmon fisheries management experience the need for a new synthesis of natural resource policy. The conceptual framework for this synthesis is to use strong scientific inference to determine what is technologically possible, and to meld this with inference about what is economically feasible and politically and legally practicable. The challenge before us is to anticipate the social conscience regarding the conservation and utilization of fisheries resources for recreational, commercial, and aesthetic purposes, and to make policy decisions consistent with the social conscience. It is an often overlooked fact that many conflicts in public policy result from overreaction to the short-term opinions of specific interests at the expense of longer term social needs and desires.

It is appropriate to consider salmonid fisheries from a global perspective. For many of us, preoccupation with local concerns results in our becoming rather provincial in our views of policy related questions. It would be beneficial for all concerned to articulate a synthesis of technology and policy as a world viewpoint. It is also appropriate to consider a diversity of professional disciplines in discussing fishery science, technology, and public policy, because the various subject areas are inseparable. Although it is convenient to partition the elements of fishery science and public policy into academic disciplines for administrative purposes, the boundaries between these disciplines become obscure when considered in the context of fishery management and utilization. The subject matter of each discipline overlaps into all of the others. Additionally, the interactions between disciplines are dynamic and continuously changing.

In exploring the interface between policy and technology, two different types of trends should be considered. The first is represented by the trends in salmon production through landings and aquaculture. The second is represented by trends in resource utilization and their influence on the policy-making environment.

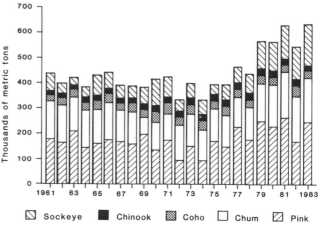

Figure 1. Landings of Pacific salmon, 1961-83.
Source: *FAO Yearbook of Fishery Statistics.*

TRENDS IN SALMON PRODUCTION

Accurate statistics on world salmon landings are difficult to compile. The following data are taken from the *Yearbook of Fishery Statistics* of the Food and Agriculture Organization of the United Nations (FAO), and provide a reasonable approximation of landings worldwide.

In recent years, the total world harvest of salmon of all species has been hovering around 600,000 mt. Pacific salmon account for approximately 94 percent of the landings and Atlantic salmon around 6 percent.

According to FAO statistics, landings of Pacific salmon have ranged from 568,000 mt in 1979 to 660,000 in 1983. (These figures do not include landings of cherry salmon of around 3,000 mt per year.) Although there is some year to year variation in both total landings and species composition, the relative constancy of the catch has led some to speculate that present levels may represent a maximum sustainable yield, at least for the Northern Hemisphere. Others argue that landings continue to reflect a modest increase over time. Landings from 1961 through 1983 are presented in Figure 1.

The species composition of 1983 landings of Pacific salmon is presented in Figure 2. The order of contribution of the five species has been conserved through the period from 1961 through 1983 with only four exceptions; chum salmon landings exceeded pink salmon landings in 1964, 1972, 1974, and 1982. Landings of masu salmon are not included in Figure 2; they accounted for less than 1

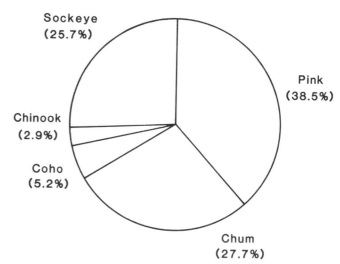

Figure 2. Species composition of 1983 landings of Pacific salmon. Landings of cherry salmon not included.
Source: *FAO Yearbook of Fishery Statistics.*

percent of Pacific salmon landings in 1983. The 1983 catch of Pacific salmon was shared by four nations: Canada, 11.1 percent; Japan, 23.4 percent; United States, 45.8 percent; and the USSR, 19.7 percent (Figure 3).

Ocean landings of Atlantic salmon ranged between 9,320 and 14,500 mt during the same period (Figure 4). These data do not include an increasing contribution of farmed Atlantic salmon (see discussion below). At least fifteen nations participated in the ocean harvest of this species during the past two decades. The proportions of the 1983 catch harvested by the several nations are presented in Figure 5.

Both Pacific and Atlantic salmon are examples of species that are harvested in commercial and recreational capture fisheries and also produced through aquaculture. However, unlike other products, the various harvest and production systems may be so closely linked as to be inseparable. One important implication of this linkage is that the prices received by both harvesters and producers are frequently fixed by international markets. An even more confounding implication is that a substantial proportion of the fish harvested in ocean fisheries result from salmon ranching where smolts are released from both public and private salmon hatcheries.

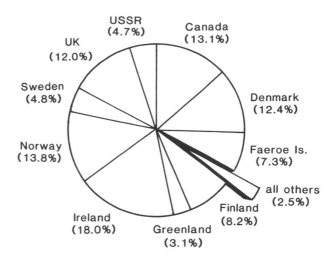

Figure 5. 1983 ocean landings of Atlantic salmon by nation.
Source: *FAO Yearbook of Fishery Statistics.*

The practice of salmon ranching dates back to the 19th century. It has been practiced on an increasingly large scale since the 1950s, coincident with the beginning of the present era of technological advancement in artificial propagation methods. Under this production concept, juveniles are produced in hatcheries, released into nature, and harvested late in the migratory phase in either offshore or terminal fisheries, sometimes in the natal rivers. In North America, hatchery production has traditionally been accomplished mainly within the public sector, but in recent years increasing proportions of the total hatchery production have been financed by the private sector.

Accurate data on hatchery releases are not available. Most estimates place current releases of juvenile salmon into the north Pacific Ocean at above four billion but less than five billion fish per year. The contribution to landings of Pacific salmon probably lies in the range of 25 to 40 percent of the total.

Future trends in ranching of Pacific and Atlantic salmon are uncertain. In view of diminishing reproductive habitat in many regions and overexploitation of many stocks, especially those captured in mixed stock fisheries, it seems likely that artificial propagation will continue to expand.

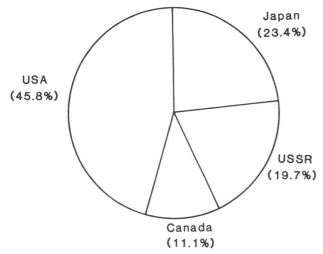

Figure 3. 1983 Pacific salmon landings by nation.
Source: *FAO Yearbook of Fishery Statistics.*

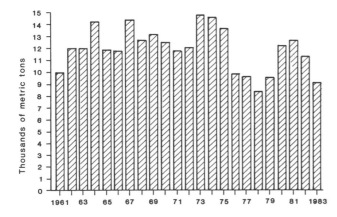

Figure 4. Ocean landings of Atlantic salmon, 1961-83. Farmed salmon production not included.
Source: *FAO Yearbook of Fishery Statistics.*

Additionally, there seems to be abundant venture capital available for investment in ranching provided the institutional climate is favorable for this type of development. On the other hand, lingering ecological and economic questions remain: What level of increased hatchery production can be attained before the ocean's carrying capacities are reached or exceeded? How much additional supply can be accommodated before markets become saturated and prices drop below levels of economic feasibility? There is some indication that present trends in supply are increasing more rapidly than demand. With some exceptions, salmon prices have not increased in proportion to inflation in recent years. Many federal, state, and provincial agencies are reevaluating the cost-effectiveness of their ranching programs in view of contemporary economic trends.

One very dramatic trend in salmon production is the increase in farmed salmon. According to the FAO, production of farmed salmon has increased approximately fivefold in the past decade, and is expected to double again by the year 2000. Production expanded from 11,800 mt in 1981 to 39,900 mt in 1985. Atlantic salmon accounts for approximately 80 percent of total production, with various species of Pacific salmon making up the remainder. Norway is the leading producer, contributing nearly 70 percent of world production of farmed salmon. Japan is the second largest producer with about 17 percent of world production. Other nations with significant and increasing productions are, listed alphabetically, Canada, Chile, Iceland, Ireland, New Zealand, the United Kingdom, and the United States.

The production of farmed Atlantic salmon and the ocean catch of this species from 1973 to 1983 are presented in Figure 6. The increase in farmed Pacific salmon is not nearly as dramatic, though the rate of increase of farmed Pacific salmon exceeds the rate of increase of ocean catch by a significant factor.

It is important to note that the recent increase in farmed salmon production is related to the quality and availability of the product. Fresh fish of very high quality can be delivered throughout the year, and the variation in quality is much lower than is found in the ocean-caught product. For this reason, farmed salmon production may continue to increase independent of supply and demand factors that influence ocean harvests.

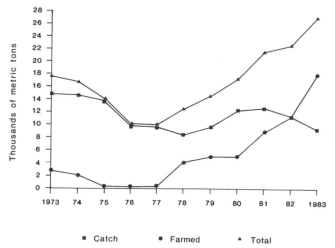

Figure 6. Ocean landings and production of farmed Atlantic salmon, 1973-83.

TRENDS IN RESOURCE UTILIZATION

Perhaps the most striking contemporary trend in resource utilization is that toward increasing complexity in the policy-making environment, which is highly visible in salmon fisheries. Consider a fictitious stock of a fictitious species of Pacific salmon. Its life history is depicted in Figure 7. In ancient times, this stock reproduced in a tributary of a pristine river system. Juveniles reared in the river and estuary, ultimately entering the Pacific Ocean. While feeding and growing at sea, they made extensive migrations, crossing present-day state, provincial, and international boundaries. In the summer of their final year of life, they returned to coastal rivers to complete their life cycle. The navigation of most was perfect; they returned to their natal river. However, a few erred to enter other streams, insuring some exchange of genetic information between stocks. Finally, following a ritualistic courtship and spawning, their carcasses fed other wildlife. Their seed, safely deposited in the interstices of the streambed, incubated through the winter and emerged in the spring to begin a new odyssey.

At some stage of the history of this stock, humans started to capture some of its members in what will be termed, for want of a better title, traditional fisheries. Initially, fish were captured for subsistence and ceremonial use by native cultures. In more recent times, nonnative settlers shared the resource, fishing for recreation as well as subsistence. Regulation of the fishery was uncomplicated at this stage. What little regulation was required was easily effected by local conservation activity. A state fish and wildlife agency handled management and enforcement. Private interests were probably limited to tribal councils, angling clubs, and sporting goods interests.

As the human population increased in the region, the demand for salmon kept pace. Inevitably, a commercial fishery developed in the lower river and estuary. The increase in human population also stimulated competitive demands on fresh water and estuarine habitats (e.g. forestry, agriculture, hydropower, urbanization, etc.). Human activities reduced escapements of adult fish for reproduction as reproductive habitat diminished. The cumulative effect of human activity was manifest in a very short time; the productivity of the stock showed a dramatic decline. But the demand for salmon continued to increase. Innovative approaches to increase the numbers of returning salmon began to appear, including construction of hatcheries.

As competition for the resource heightened, it was only a matter of time until the interception of fish in the ocean became a common practice. Commercial and recreational fisheries near the natal rivers appeared first, but the harvest spread to more and more distant waters. Ultimately, nationals of foreign nations began to harvest portions of the stock in international waters.

The flourishing economic development in the watershed and in the offshore fisheries was attended by a dramatic increase in institutional complexity. All of the various activities required research, assessment, coordination and, in some cases, licensing. The agencies and institutions involved seemed to grow in geometric proportion to the number of user groups. At this stage, the public agencies and private interests influencing policy decisions would total at least several dozen and would include, for example, state departments of agriculture, forestry, environmental quality, and water resources; federal agencies such as the Bureau of Land Management and the U.S. Forest Service; and private interests ranging from the tourist industry through the construction industry to seafood harvesters, processors, and brokers. The trend toward increasing institutional complexity with time has become a familiar theme in

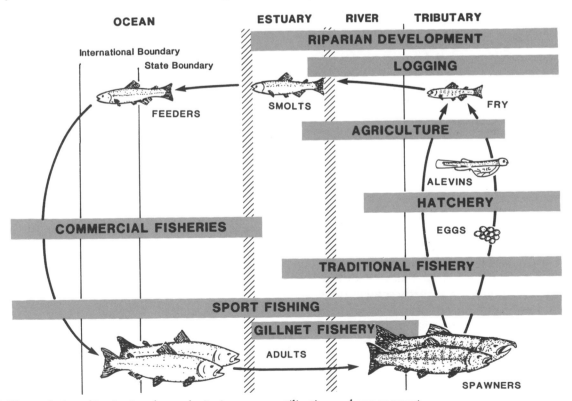

Figure 7. The evolution of institutional complexity in resource utilization and management.

contemporary salmon fisheries, and underscores the need for new approaches to formulating fisheries management policy.

FUTURE PERSPECTIVES

In view of contemporary trends in world salmon production, one may be tempted to speculate about future production. However, one must approach such predictions with caution because future production is contingent upon issues that are as yet unresolved.

From a purely technical perspective, there is no apparent reason that world salmon production could not increase substantially in the years ahead. Although significant increases in natural production seem improbable, the technology presently exists to increase production through both salmon ranching and salmon farming, and technological advances are occurring almost daily. However, increased production does not necessarily follow technological advance. Future trends in salmon production are less likely to be determined by technical considerations than by marketing factors and a plethora of institutional concerns, including questions of management policy and international law that have not been formally or adequately addressed. It is a widely held view that the future of world salmon production is closely linked to the need for better mechanisms for making policy decisions. Those holding this view cite two confounding problems with traditional approaches to formulating natural resource policy.

First, public policy decisions are typically based on technical inference and fuel technical innovation. As policies to conserve natural resources evolve and are implemented, new technologies arise that improve the efficiency of resource utilization. Equilibrium is never quite achieved. Second, the development of natural resource policy is typically after the fact; policy is formulated in response to conflict.

Recent diplomatic and legislative actions may be construed to indicate that a new synthesis of natural resource policy is already under way. In order to increase salmon production for the benefit of society, and probably to even maintain a *status quo*, this synthesis must continue and be strengthened in at least two ways.

First, policy must become proactive as opposed to reactive. It must anticipate societal needs rather than merely respond to the concerns of specific interests. Second, insofar as policy is based on technical inference, emphasis must be placed on strengthening the nature of the inference. Weak inference, and specifically weak statistical inference based on empirical records, places the credibility of public policy in jeopardy.

The development of a world view on salmon resource policy is a matter of considerable urgency. In its absence, the potential benefits of technological advance may be wasted. This loss is tantamount to depriving society of the full opportunity to enjoy and utilize salmon resources for commercial, recreational, nutritional, and aesthetic benefit.

An Interpretation of Salmon Production Trends

William F. Royce

Professor Emeritus
School of Fisheries
University of Washington

When the first elected governor of Alaska addressed the Joint Assembly of the First Alaska State Legislature in January 1960 he stated, "On January 1 of this year, Alaska's Department of Fish and Game was handed the depleted remnants of what was once a rich and prolific fishery. From a peak of three-quarters of a billion pounds in 1936, production dropped in 1959 to its lowest in 60 years."[1] He was referring primarily to the salmon fishery, which had produced 85 percent of that record catch, and which had declined in the 1959 season to only 23 percent of its 1936 record.[2]

The State of Alaska was taking over management of the most valuable salmon resource in the world at a time of great tension over harvest management. The International North Pacific Fisheries Commission had ended a 5-year study period in 1958, after which it was to decide annually whether salmon, halibut, and herring stocks in North America should receive protection from Japanese high seas fishing. In addition, massive research efforts on salmon in Canada and the United States had been launched during the 1950s to improve salmon management.[3]

After the disastrous low catch of 1959, annual salmon production by the United States recovered

13

modestly, only to decline to even lower levels in the mid-1970s. Since then, however, the increase in salmon production has been extraordinary; the fishery has become richer and more prolific than ever in its recorded history. The reasons for this surge are not fully understood but they must include factors of special significance to salmon management.

The decline and restoration of the Alaskan salmon fisheries to record levels, after intensive exploitation for a major part of the last century, deserves critical attention, especially because good management appears to have contributed significantly to recovery of the resource. I propose to examine long-term trends in several salmon fisheries and hypothesize biological explanations for a resurgence of the kind that happened in Alaska.

TOTAL NORTH PACIFIC PRODUCTION

After the end of World War II, the Japanese rebuilt their fishing capability, and Asian salmon production from wild stocks surged to nearly 300,000 mt in the late 1950s. Since then it has gradually declined to a level less than 150,000 mt annually. By contrast, North American production, after relatively unhampered fishing during the war, fell to a level of less than 200,000 mt in the late 1950s, surpassed Asian production in the early 1960s, fell to another low in the early 1970s, but since has surged to record levels above 350,000 mt in the early 1980s (Figure 1).[4] Since the mid-1970s, all-time or at least 50-year record production of sockeye, pink, chum, and coho has occurred in from one to several years in each of the major Alaskan fishing districts; southeastern, central, and western. Yukon-Kuskokwim production has also been excellent.[5]

These figures on wild salmon production do not include the hatchery enhancement which has exploded in Asia to more than 100,000 mt annually in the early 1980s, and increased in Alaska and Canada to about 30,000 mt by the mid-1980s.

Canadian salmon production has usually ranged between one-third and two-thirds as much as that from the United States, and the fishery management systems of the two countries have been similar. But recent Canadian production has remained about average—not surging as has production from most districts in Alaska.

Early Alaskan production

The Alaskan fisheries were under federal management from the purchase of Alaska in 1867 through 1959. Throughout this period, the salmon fisheries were the most important resource of the area. Commercial salmon production began in the 1880s, rose quite steadily to a peak of about 235,000 mt in 1918, then became volatile, reaching new peaks of about 250,000 mt in 1934 and 290,000 mt in 1936. After these seven decades of exploitation came an almost steady decline through the 1950s to a level below 100,000 mt (Figure 2).[6]

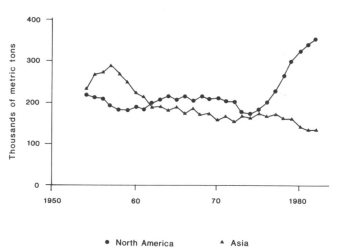

● North America   ▲ Asia

Figure 1. Five-year running average of salmon harvest in North America and Asia.

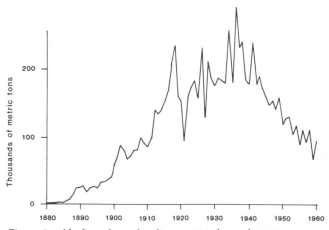

Figure 2. Alaska salmon landings, 1880 through 1960.
Source: Lyles (2).

Early Columbia River production

Commercial production of salmon in the Columbia River started about 20 years before it started in Alaska, and reached a peak of 20,000 mt in the early 1880s—about the time that Alaskan production began. During the 50 years after this first peak, production was volatile, although averaging about 15,000 mt annually (Figure 3).[7] Almost all of the recorded production until the first peak was chinook salmon, after which coho, sockeye, steelhead, and chum were included in increasing proportions.

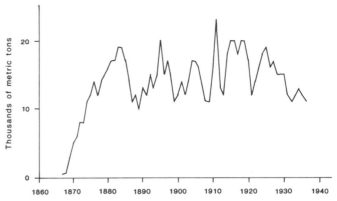

Figure 3. Columbia River salmon landings, 1867 through 1936. Source: Craig and Hacker (7).

Early Canadian production

Early trends in total British Columbian salmon runs were affected to a major extent by environmental factors. The fishery for sockeye returning to the Fraser River developed first, but this was greatly reduced by the 1913 landslide at Hell's Gate in the Fraser River canyon. The odd-year runs of pink salmon also returned predominantly to the Fraser River and nearby streams, but before 1910 large quantities of these were caught incidentally along with the preferred sockeye and discarded.[8] The sockeye and the odd-year pinks dominated runs in the Fraser River and both were heavily fished by U.S. fishers, whose catches came also from the Puget Sound runs.

The data on even-year pink salmon are useful to evaluate long-term trends in natural production in central and northern British Columbia. Canned packs of even-year pinks rose quite steadily from almost none in the first decade of the century to a peak of more than one million cases (about 32,000 mt) in 1930 then collapsed and fluctuated from the 1930s through the 1950s (Figure 4). This volatility continued into the 1980s even though peaks occurred again in 1962 and 1966.[9]

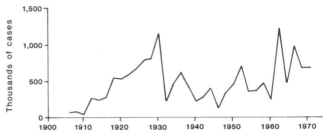

Figure 4. British Columbia canned pink salmon production for even years, 1906 through 1970. Data from Pacific Fisherman Yearbooks.

THE PHASES OF EARLY PRODUCTION TRENDS

The long-term trends in catches from Alaska, the Columbia River, and British Columbia show a remarkably similar phasing: a relatively steady increase to an initial peak after two or three decades, followed by a period of pronounced volatility for two or three decades at a significantly lower average level than the first peak, and, finally, a decline, with somewhat less volatility in the Alaskan and Columbia River catches than in those of British Columbia.

This similar phasing is more remarkable because of the major differences in the species involved. Alaskan production involved all five species of North American salmon with a cyclic mixture of 2-year pinks and 4- to 6-year sockeye, chums, coho, and chinook. Columbia River production was predominantly chinook, augmented after the major peak by increasing proportions of coho, chum, sockeye, and steelhead trout. The even-year pink production in British Columbia was, of course, a single species with a uniform 2-year life cycle.

The common factor in the first phase of each of these major salmon fisheries appears to be the rapid increase in catches associated with expanding markets in which the choice chinook and sockeye were preferred. As catches of these species peaked, fishing turned to the lower value coho, pinks, and chum. The timing of the first peaks was consistent with the use first of the preferred species in areas closest to population centers. The first peak in the Columbia River chinook catches was in 1883-84, in the five Alaskan species in 1918, and in the even-year pinks of British Columbia in 1930.

SALMON MANAGEMENT

Until the 1950s, government management of North American salmon fisheries had virtually no basis in research on salmon populations or on the need to regulate fishing according to genetic units. Early regulations were simply based on the understanding that salmon returned to their home streams and had to have access to their spawning areas. Restraints on fishing were few, partly because of widespread dependence on salmon hatcheries, and partly because regulations were almost entirely promulgated by legislatures influenced by business needs.[10]

The role of population research in salmon management emerged with the formation by Canada and the United States of the International Pacific Salmon Fisheries Commission to study the salmon runs of the Fraser River. A treaty was negotiated in 1930 but blocked by political opposition from the State of Washington until 1937 when it was finally ratified. This led to alleviation of the blockage at Hell's Gate in the Fraser River canyon, as well as research on regulation of each major spawning unit of the sockeye. After public acceptance of regulations based on research, the treaty was modified in 1957 to give the Commission responsibility for recommending regulation of pink salmon as well as sockeye.[11]

The next major step in applying population research to salmon management was instigated by the Alaska salmon industry, alarmed by the 50 percent decline in Alaskan production between the mid-1930s and the mid-1940s. Dr. W. F. Thompson, who had shown the importance of research when he was Director of the International Pacific Salmon Fisheries Commission, was encouraged to apply similar approaches in Alaska. With this industry support, he formed the Fisheries Research Institute at the University of Washington in 1947, and began the first significant population research on salmon in Alaska.

Another international agreement was required to really instigate extensive research on the salmon populations up and down the coast. The International North Pacific Fisheries Commission originated from fears that Japanese fishing off the North American coast could decimate populations of salmon, halibut, and herring. A treaty based on the concept that fisheries on these stocks would be regulated for maximum sustained productivity was developed by the United States, Canada, and Japan. Both Canada and the United States began massive research programs in the mid-1950s and both forbade net fishing for salmon by domestic fisheries outside a "surf line" in order to strengthen the regulation of fishing on each genetic unit.

Perhaps the lack of salmon management before the 1950s is best indicated by personnel and appropriations in Alaska, where a large majority of the North American catch was taken after the early 1900s. Total law enforcement personnel increased from about twenty in 1920 to a peak of about 320 in 1930 (mostly stream guards), and then declined to a level of about 120 in the mid-1940s.[12]

Another indication is the increasingly later date when 75 percent of the catch was taken in the pink salmon fisheries of Alaska between 1910 and 1940. The management practice was to allow full-scale fishing until a permitted total was obtained, and then to close the fishery to allow escapement. This practice ignored the fact that the later runs spawned in different places than the early; they were different genetic units. This was a positive link between heavy fishing and severely declining runs in the early parts of the seasons.[13]

Other reports indicate catches of very large proportions of salmon runs in the Columbia River. Counts of the first runs to use the fishways in 1938 after the closure of the Bonneville Dam on the Columbia River indicated that only one out of four to seven fish escaped the river fishery—and this excludes the effects of ocean catches, sport catches in the river, or subsistence catches on the spawning grounds.[14]

HYPOTHESES

During the earliest stages of the developing fisheries the catches must have been limited by the fishing and processing facilities and markets. These, however, expanded rapidly, and Columbia River production reached its first peak in only about two decades. These initial peaks have been equalled only occasionally later in all three areas, except for Alaska's sustained records of the early 1980s. The first peaks that were reached so smoothly appear to have been approximate indicators of the maximum level that might have been sustained with proper management and without major environmental change.

The catches after the first peaks of the essentially unregulated fisheries appear superficially to resemble the cycles of the longer lived species such as sockeye or chinook. But these cycles are irregular and their length is unrelated to the age of dominant species. Thus the peaks must have involved mixtures of damaged runs and others that accidentally benefited from reduced competition in their niches or favorable environmental conditions.

The consequences of virtually unregulated fishing must have been capture of radically different proportions of individual stocks of widely varying abundance, and decimation or virtual extinction of some of them. This almost surely reduced the adaptability of the runs as a whole to changing fishery and environmental conditions.[15]

The decline in total catches in Alaska after the mid-1930s appears to have been a consequence of continuing depletion of individual stocks—probably those that had been most resilient during the period of volatility. Such might also be the case in the Columbia River after 1920. At this time numerous tributary dams had been built but none of the major main stream dams. The principal runs were of chinooks, which are big river spawners, and hence appear likely to have been decimated much more by uncontrolled fishing than by the dams.

After such depletion, how can the record production in Alaska during the 1980s be explained? It involved many runs of all species, except chinooks, in all major fishing regions of Alaska, and it has been sustained for several consecutive years—a surge not previously appearing during any of the periods of volatility discussed above. Clearly it has not been unique to a species, an area, or a year.

Perhaps it has been associated with fortuitous weather that improved both freshwater and ocean conditions—a reversal of the weather of the mid-1970s. If so, it is surprising that similar surges have not occurred in the previous history of the Alaska fisheries or in the Asian fisheries of the 1980s. I suggest that while weather may have been a factor,

it could not have produced such a surge without restoration of depleted stocks. Alaskan salmon management has become increasingly competent since the late 1950s to protect depleted stocks from overfishing, particularly through improved ability to predict sizes of returning runs.

Moreover, there may also have been sufficient time for depleted stocks to become at least partially restored as a result of fish straying into unoccupied niches, and having their progeny protected. Twenty years is about the time required for emergence of behaviorally modified genotypes in quantity, if we judge by the results of salmon transplantations, salmon occupation of stream areas opened up by fishways or receding glaciers, and by the number of generations required for behavioral modifications in short-lived laboratory animals.

The implication of this look at the long-term history of some salmon fishing regions is that uncontrolled fishing on mixed stocks can destroy a large proportion and hence cause major loss of productivity. But it appears that runs can be restored to unoccupied niches by better management, by transplantation, or by hatchery introductions.

These are not new concepts. They are an elaboration of the forecast that I made in 1963 when the level of Alaskan production was about 100,000 mt and after the lack of effective management had been detailed: "I believe that it is technically possible and economically practical to double the production of Alaskan salmon . . . we must protect the weak runs and harvest the surplus from the big ones. Some weak runs require total protection while the big runs can be harvested at a rate of 85 to 90 percent. A uniform catch rate such as the 50 percent rate formerly imposed by the Federal Government is now recognized as nonsense."[16]

LITERATURE CITED

1. R. A. Cooley, *Politics and Conservation: The Decline of the Alaska Salmon.* Harper and Row, New York, 1963. See p. xix.
2. C. H. Lyles, Fishery Statistics of the United States, 1966, Stat. Digest 60, Bur. Comm. Fish., U.S. Fish Wildlife Serv., (1968). See pp. 515-526, Historical Fishery Statistics, Alaskan Salmon Fishery.
3. R. I. Jackson and W. F. Royce, *Ocean Forum: An Interpretative History of the International North Pacific Fisheries Commission.* Fishing News Books, Farnham, England, 1986. See Chapter 3.

4. Jackson and Royce, *Ocean Forum*, p. 174.

5. *The Fishermen's News*, Pacific Fisheries Review, March 1986.

6. Lyles, Fishery Statistics.

7. J. A. Craig and R. L. Hacker, The history and development of the fisheries of the Columbia River. U. S. Dept. Interior, *Bull. Bur. Fish.* (32):133-216, 1940.

8. G. A. Rounsefell and G. B. Kelez, The salmon and salmon fisheries of Swiftsure Bank, Puget Sound, and the Fraser River. U. S. Dept. Commerce, *Bull. Bur. Fish.* (27):693-823, 1938. See p. 809.

9. *Pacific Fisherman Yearbook*, 1966, and *Pacific Fisheries Review* subsequently.

10. P. A. Larkin, Management of Pacific salmon off North America. Pages 223-236 in N. G. Benson, ed., *A Century of Fisheries in North America.* Amer. Fish. Soc., Washington DC, Sp. Publ. 7, 1970.

11. International Pacific Salmon Fisheries Commission, Reports 1937 ff.

12. W. F. Royce, Pink salmon fluctuations in Alaska. Pages 15-33 in N. J. Wilimovsky, *Symposium on Pink Salmon*, H. R. MacMillan Lectures in Fisheries, Inst. of Fisheries, Univ. of British Columbia, Vancouver, 1962.

13. Ibid.

14. W. H. Rich, The salmon runs of the Columbia River in 1938. U. S. Dept. Interior, *Bull. Bur. Fish.* 50(37):103-147, 1942.

15. W. E. Ricker, Hereditary and environmental factors affecting certain salmonid populations. Pages 27-160 in R. C. Simon and P. A. Larkin, *The Stock Concept in Pacific Salmon*, H. R. MacMillan Lectures in Fisheries, Inst. of Fisheries, Univ. of British Columbia, Vancouver, 1962. See Epilogue pp. 145-150.

16. W. F. Royce, Prospects for Alaska salmon. Fisheries Research Institute, Circular 203, 5 pp. Speech presented at Fourth Annual State Convention, Alaska State Chamber of Commerce, Juneau, Oct. 19, 1963.

The Laws Covering Salmon Ranching from the Salmon's Point of View

Alfred A. Hampson

*Hampson, Bayles, Murphy and Stiner
Portland, Oregon*

In order to deal with this subject, I must endow my mythical ranch salmon with a certain degree of anthropomorphism so that it can think like a middle-aged lawyer or have the middle-aged lawyer look at this whole problem from a fishy point of view. I will probably do both and it may be difficult to distinguish between the two points of view, the fish as a lawyer or the lawyer acting like a fish.

The salmon from whose point of view I am looking at the law is a ranch salmon. It is hatched in a hatchery and reared in a pond owned by an aquaculturist or salmon rancher. It is then released to spend the next 2 to 5 years in the pelagic ocean and then to return to its release site. There, the eggs or milt will be taken to procreate other salmon to carry on the race and the business of salmon ranching. Its carcass will be sold.

From the salmon's point of view the problems that are of greatest concern are: "Who, if anyone, owns me?" and: "Who may fish for me, when, and under what conditions?" The salmon has asked me to examine its rights under common law, legislative law, and state, national, and international law. It also asked me to make a distinction between laws and rights that are theoretical and those that are practical.

In order to explain these various approaches to the ranch salmon it is first necessary to explain to it a little property law. The right to own property is really the right to defend its possession either by yourself in primitive societies or by the state in modern times. The state says that if you buy something it will send its police to keep someone from stealing it from you, or if you sell it and the buyer fails to pay the state will open up its courts so you can sue the buyer and enforce payment. Property rights can expand under edict of the state. Examples include new rights in intellectual property. Property rights can also contract. Zoning imposed on previously unzoned land would be an example of this.

With this sound understanding of property law, let us look at the ranch salmon's ownership and therefore its rights under common law. Initially common law defined three classes of animals, and all creatures had to fit into one of the categories.

The first classification describes domestic animals (*domitae naturae*) such as horses, cows, and pigs. Ownership attaches to the animal and the owner has an absolute property right to it. If the cow escapes, it is still owned by its owner, who may come and recapture it from a neighbor's field. A ranch salmon that is reared in captivity and released to range in the pelagic ocean and then to return to the release facility to be captured by the salmon rancher bears little resemblance to the Holstein that has broken down the fence and is enjoying the neighbor's clover. Nor does it resemble a pig that has escaped and become feral. Such a pig has no intent to return to its original sty, but if it did return it would again be a domestic animal. It seems unlikely that a ranch salmon would be defined by common law as a *domita natura*.

The second classification describes wild animals (*ferae naturae*). These are creatures which are not owned by a person until the animal is reduced to the captor's absolute possession with no means of escape. The mallard in the sky, the mule deer in the forest, and the coyote on the bluff are wild and free. They belong to no one until captured. Over the years the courts have had many cases in which they sought to determine just exactly what "captured" meant. Unfortunately from the duck's or the salmon's point of view the cases were between two claimants over a dead wild creature. One example is a case where a plaintiff was in the process of closing his seine when the extremely brave defendant entered the plaintiff's net, surrounded the fish with his own net and captured them. The plaintiff's suit was dismissed because at the time that the defendant took the fish the plaintiff had not yet reduced the fish to his possession because they could have escaped from the unclosed seine.

Returning to our ranch salmon, it would seem that the smolt, which is owned absolutely by the salmon rancher when it is in the rearing pond, becomes a *fera natura* once it is released to forage in the ocean. When it is hooked by a commercial or sports fisher and reduced to his or her possession, then it becomes the absolute property of that fisher. I suppose if the salmon is eaten by another fish it would cease to be a *fera natura*, and that fish would have absolute ownership of the poor salmon. When the fish swims through the weir, it again becomes the property of the salmon rancher. If it swims back through the opening in the weir or over the top of the weir in a freshet, it again becomes a *fera natura*, free to be reduced to possession and ownership by any fisher who has the skills and the luck to do so.

The common law has a third category in which the anadromous ranch salmon might fall, the category of *animus revertendi*. This is a creature which was owned, released into the wild where it was free to become wild but, in fact, returned to the owner. Racing pigeons and hunting hawks are such creatures. Blackstone in his *Commentaries* said: "For my tame hawk that is pursuing his quarry in my presence, though he is at liberty to go where he pleases, he is nevertheless my property; for he hath *animus revertendi.*"

Obviously there is a difference in the length of time a pigeon or a hawk is free "to go where he pleases" before returning to the cote or the owner's arm and the period of time the ranch salmon has between release and return. Maybe the period of time is too long for the salmon to be considered *animus revertendi* and it should be considered as a *fera natura*. But, on the other hand, the instinct of the anadromous salmon to return to the stream of its birth predates humanity and certainly predates the imposition of a human's will on pigeons for the purposes of racing or on hawks for the purposes of hunting. Ichthyologists know that if the salmon can come home to its stream it will do so.

There may be a fourth category or possible merely a subcategory of common law. It is new and sometimes referred to as electronic common law. If a salmon were marked by a distinctive fin-clipping or by a coded-wire tag, would it retain the characteristics of a domestic animal and stay private property? There is a similarity to the steer which has been branded and turned out into the public domain to graze, fatten, and then to be rounded up. The ranch salmon do not even have to be rounded up since they are self-herding.

Obviously, the common law and Blackstone's *Commentaries* do not give us a clear solution. To try to find one, let us examine the traditional and legislatively created types of property. There is private property, public property, and common property. In the course in common law which we and the salmon just completed we learned that the sovereign, the state, or the legislature has denominated certain interests as private property and will defend them. There are those who urge that, to ensure the success of salmon ranching, the state should create a right in the smolt which was raised as private property in a private hatchery. That is, that the smolt be clothed with the title of private property and protected by the state from the time of release to the time of recapture no matter where it swims or how long it stays at sea.

Public property is similar to private property except that the owner is the state or one of its agencies. The White House is owned by the United States of America, not by the people of the United States of America. A citizen cannot use it or even visit it at will, but only when the National Park Service permits, and then only by complying with the Park Service rules and conditions. The Forest Service owns the trees in the national forests and may sell them to private loggers in the same way that private timber owners might sell their trees. The Bureau of Land Management leases public land for the drilling for oil or the mining of coal. These are examples of public property which, in fact, is treated like private property.

In certain jurisdictions ranch salmon and, in fact, all salmon are called public property. That is, they are owned by an agency such as the Oregon Department of Fish and Wildlife or the California Department of Fish and Game. I do not think this is good law—or at least good English. A wild salmon which spawns in the wild without human interference as its ancestors have done for millions of years and its offspring that have gone to sea and returned to the same wild reach of the river are not made public property merely by having the

legislature say so. Such a salmon is truly a *fera natura*, and the legislature has not reduced it to effective possession.

What about a salmon raised in a state hatchery for the purpose of increasing the stock of fish available to be caught by commercial and sports fishers? These fish are caught in Oregon, Washington, California, British Columbia, and Alaska waters. Does the California legislature determine the property rights in a fish that was hatched in a California hatchery but is caught off Alaska?

It is my opinion that a hatchery fish, either from a state hatchery or a private hatchery, when it is released becomes a *fera natura*, and that the state agency has given up or is unable to assert a property interest in the salmon once it has been released. To be sure, the state can and does say who may fish for a salmon within the waters controlled by the state and when, where, and with what gear they may do so. But the state could not sell the uncaught fish as public property as the Forest Service sells an unfelled tree in a timber sale. The most that the state can sell is a license for the fisher, commercial or sports, to fish for a salmon, but a salmon becomes the private property of any licensed fisher who catches it and reduces it to his or her effective possession.

Common property rights are rights to a natural resource that is open to a class of users whose rights are coequal. The commons in medieval England or the village green in Massachusetts were or are available to be used by all the qualifying people equally, and none may be excluded by the others. If the green is so held that all of the inhabitants of the village may use it, the villagers hold it in common, as common property. This is true whether the population of the village is five hundred or five thousand. If the salmon in the ocean, excluding for a moment ranch salmon, may be fished for by anyone and when caught become the private property of the successful fisher, then that body of uncaught salmon is common property.

In addressing ranch salmon, Alaska law deals with this problem in a clear straightforward manner. Alaska revised statutes state: "Salmon reared by private nonprofit hatcheries are declared to be a common property Alaskan natural resource from the time they are released into the natural waters of the state until the time that they segregate from naturally occurring stocks at the specific location designated in the permit by the department."

Geography may affect the ownership of the salmon and will certainly affect its relationship with fishers. As our intellectual and contemplative ranch smolt swims west and gradually grows into a salmon of commercial value it might ask itself "Who owns me?" and "By what laws is that ownership determined?" and, much more to the point, "What laws govern the conduct of those whose interests might be inimical to my health, happiness, and longevity?" Whether our salmon is denominated as public property or common property or is a wild creature, a *fera natura*, or something else is immaterial. There are lines to the west that affect it. The 3-mile limit was, since the beginning of our nation, treated as an extension of the sovereignty of the state. The state determined who could fish for salmon in what manner and up to what limits in its rivers and streams; and, too, it could do so out to the 3-mile limit, known as the territorial zone.

In 1947 in *U.S. v. California* 332 U.S. 19, in a suit to determine whether the State of California or the United States owned the offshore oil, the Supreme Court held that the United States owned the territorial sea, out to the 3-mile limit. However, Congress subsequently passed the Submerged Lands Act of 1953 which confirmed the titles in the states to the lands beneath the navigable waters within state boundaries and to the natural resources within such lands and waters. The act defined fish as a natural resource. Thus, while a state may not own the fish, as it cannot control it or reduce it to possession, it does control it inasmuch as it can keep its own citizens or citizens of other states or countries from trying to control it. The "contiguous zone" extends from the territorial zone an additional 9 miles, creating the 12-mile limit. The Submerged Lands Act gives title to the land under the contiguous zone to the federal government. In 1966 an exclusive fisheries zone was set up in the contiguous zone and fishing vessels from foreign nations were prohibited from fishing in that zone. That act also specifically denied any intention to extend the jurisdiction of the states to the "natural resources beneath and in the waters within the fisheries zone."

In 1976 Congress passed the Fisheries Conservation and Management Act that extended the exclusive fisheries zone to 200 miles from shore, thus creating the 200-mile limit. Our ranch salmon should know that so long as it is within the 200-mile zone it doesn't have to worry about foreign fishing boats. The 1976 law also prohibits foreign vessels from fishing beyond the 200-mile limit for anadromous fish originating in the U.S. unless they are in foreign waters. The United States has therefore claimed ownership to salmon hatched in the U.S. that are swimming in international waters, beyond the 200-mile zone. It would seem that this claim of ownership may well run counter to international law and from a legal point of view, except as controlled by the treaty where the U.S. is a signatory, would be difficult to assert. In addition, from a practical point of view the law enforcers of the United States will be hard pressed to say whether an anadromous fish caught 250 miles off the Washington coast is a wild or hatchery fish from one of the western states, from British Columbia, or from Russia. Even though the United States has asserted this claim, our ranch salmon, when it is beyond the 200-mile limit, would be well advised to realize there are no laws to protect it; and thus it is owned by no one and no one can limit those who can fish for it. Additionally, if our salmon swims within the 200-mile limit its relationship with those who would catch it changes should it enter waters controlled by Canada.

How is the perplexed ranch salmon to deal with these confused and confusing laws and assertions that surround it? Worse, how is the confused salmon rancher to deal with them? What is needed is a policy that addresses the problems of the salmon rancher as they now exist, without reference to the common law or the terms of private, public, or common property.

A major issue is: Should a common resource be used for private gain? Would it be a wise public policy to permit salmon ranchers to release their fish to pasture in the ocean, which is a common resource, whether it is controlled by a contiguous state, by the United States by fiat or treaty, or by no one? In trying to determine what would constitute a wise public policy, we should be mindful of the mistakes which the United States and the states have made in the past in permitting the public or common lands to be appropriated for private gain. Any policy decision should, insofar as possible, be based on the facts as we actually know them and not on mere assertions.

The problems, as suggested by Bowden (1981) in his excellent book *Coastal Aquaculture Law and Policy: A Case Study of California*, are (1) the carrying capacity of the ocean; (2) the competition between the salmon rancher and the commercial fisher; (3) control of the new salmon ranching industry; and (4) who can best control the resource.

In discussing the carrying capacity of the ocean, Mr. Bowden discusses the problems of managing common property. He states: "Economic logic dictates that when a natural resource is available to multiple users without limitation on the use each may make of it, the self-interest of each user will lead to the destruction of the resource." When the natural resource, whether it is the range with multiple ranchers entitled to graze their cattle on it or the ocean with unlimited fishers, begins to become depleted, all will increase their efforts to exploit it as fully as possible until the resource is destroyed. Mr. Bowden calls this "the last anchovy argument"; i.e., the salmon ranchers will pump more and more salmon into the ocean until eventually one of the ranch salmon will eat the last anchovy.

That the carrying capacity of the ocean can be exceeded seems unlikely when one thinks of the huge numbers of salmon that the Pacific once carried. Mr. Bowden points out that there were once 6,000 miles of streams in the central valleys of California for the salmon to spawn. Now there are 500 miles. The situation is similar in the Columbia River drainage where miles of spawning gravel have been flooded out or completely cut off, as in the case of spawning grounds above Grand Coulee or Hell's Canyon. One factor that is often overlooked when the ocean is compared to the great grasslands of the West, which were overgrazed and destroyed, is that the cattle ranchers had very largely exterminated predators. The wolves and the coyotes were killed as rapidly as possible. In Africa, where the carnivores follow the huge migrations of grasseaters, the size of the herds is kept in balance and the grazing resource is maintained. It is not possible for humans to destroy the ranch salmon's predators, since not only do other species eat the salmon but the salmon eat each other.

In a rather convoluted way commercial fishers and salmon ranchers are not competing for the same resource. Fishers are hunting for wild or hatchery fish. Salmon ranchers seek an ocean in which to pasture their fish. They do not seek any of the wild fish or any hatchery fish other than their own. This analysis does not go quite far enough, however, because the wild and hatchery fish from a state hatchery pasture in the same ocean where the ranch salmon pasture, and are hence competitors for the same resources.

If the carrying capacity of the ocean is finite, and not self-regulating, then the possibility of its being exceeded must be addressed. It is necessary to consider the motives of the advocates of various positions on the subject. Commercial fishers have consistently urged that more and larger hatcheries be placed in the Columbia River drainage. Do the state and federally funded hatcheries create no impact on the number of anadromous fish that the ocean can carry, while private hatcheries do? Perhaps commercial fishers are motivated by the fear of competition and not seriously concerned about how many fish the ocean can carry.

The determination of carrying capacity must be based upon the best evidence available. There are certain things that we do know, and one of these is that we know very little. Where do the salmon go? What do they eat? Who eats them and how do they fare when the cycles of their food and predators change? Do some salmon, some of their predators, and some of their prey have wider tolerances for changes in the temperature of the ocean than others? We can hesitate at the brink of the ocean forever awaiting more and better evidence. We should act now, and if we are to create a legal system that aids the samon rancher we must also create a system that will permit us to change our decision if it proves to be wrong.

There is a conflict or an apparent conflict between the commercial fisher and the salmon rancher. There are emotional arguments on each side. One argument is that salmon ranching is the wave of the future. It is inevitable. It will create new jobs, new wealth, and cheaper food. On the other hand are arguments that ranching will destroy the small independent fishers and the towns along the coast, and once more America will lose a portion of its independence as it moves toward greater concentration of wealth. Probably each argument is in part correct and in part incorrect. The commercial fisher hunts the wild fish and the fish from state operated hatcheries, and would be free to hunt ranch salmon also. It would seem that the interests of the commercial fisher and the salmon rancher are not mutually exclusive.

Mr. Bowden raises the problem of how to control the salmon ranchers if they are given the right to pasture their salmon in the ocean, the common property of all of us. A user of common property has a tendency to try to exclude others from it. For instance, it is not hard to imagine the salmon ranchers seeking to have the fishing seasons changed to their advantage or having the hair seals killed.

In recognizing these possibilities and the creation of "a new class of robber barons," to quote Mr. Bowden, there already exists a group of exploiters of the common who contribute very little. Commercial fishers convert the common property—the wild and hatchery fish from a state hatchery—to their own private property, just as salmon ranchers seek to convert the pasture in the ocean to their private benefit. Neither seems to be a greater or lesser robber baron except that one group has been doing it longer.

The fears of the sports fisher appear to be based upon these questions: Who is to control the ocean? What would be the effects of ocean ranching on the composition of the genetic pool of natural spawning stocks? The sports fisher appears to be confused by three beliefs that may or may not be true but are strongly held: (1) wild is best; nature knows what it is doing; (2) any hatchery fish that strays can only debase the wild fish and cannot contribute the phenomena of hybrid vitality; and (3) interbreeding is bad, except possibly in a discrete drainage where no natural straying occurs. We should try to examine fish genetics in a factual and nonemotional way and then do what is best for the salmon.

The problem of how to restrain those who seek to control this common property is difficult, and how to control the salmon ranchers in particular is even more difficult. Two possibilities are: (1) have the state be the entrepreneur and operate the salmon ranches; and (2) have the salmon ranchers be given a free hand to control and exploit the ocean. A third alternative is to have an honest and sensitive bureaucracy, which is beholden to none of the users of the common property, seek to control the varying interests. If each group is forceful and able to express its point of view the possibility of the controlling agency being captured by one group to the detriment of the others or the public is lessened.

Appropriate priorities must be set in determining policy. The first priority must be the salmon and its survival, then what is good for the world food supply, followed by what is good for the United States and the states facing the Pacific. The last consideration is what is good for the salmon rancher or the commercial or sports fisher.

The policy must create certainty so that salmon ranchers know where they stand and can assess the risks of their ventures (such as marine survival, freshets at the recapture site, and market conditions) without having to worry about their legal position. The policy must be fair. To quote Mr. Bowden, "It seems unreasonable, for example, to treat the salmon rancher like a poacher on the commons if we are unwilling to apply the same standard to the commercial fisherman. Surely equity demands that all users be treated alike."

I believe that legislation should be adopted somewhat similar to Alaska's or the suggested model act for Nova Scotia as set forth in Wildsmith (1982). The legislation should pronounce that the ocean is a common resource and that a ranch salmon is common property which, when caught by commercial or sports fishers, becomes their private property, as it would become the salmon rancher's private property when it returned through the weir at the rancher's recapture facility.

Again following Nova Scotia's model act and the Alaska law, the legislation should provide that neither commercial nor sports fishers may fish within a certain distance of a recapture facility. This would put a floor under the risks that the salmon rancher would assume from commercial and sports fishers, but leave open the possibility of a larger return under the right circumstances. Such legislation would do much to clarify an area that is a morass of legal principles into which salmon ranching does not fit.

But what about all of this from the salmon's point of view? I would add to the legislation that a ranch salmon, from the time of release until the time of recapture, is a *fera natura*. Then it will know for sure that it belongs to no one, no state, and no nation, no matter what they gainsay.

LITERATURE CITED

Bowden, G. 1981. *Coastal Aquaculture Law and Policy: A Case Study of California.* Westview Press, Boulder. 241 p.

Wildsmith, B. H. 1982. *Aquaculture: The Legal Framework.* Edmond Montgomery, Ltd.

Salmonid Programs and Public Policy in Japan

Yoshio Nasaka

Economic Fisheries Specialist
American Embassy
Tokyo, Japan

JAPAN'S SALMON SUPPLY

Japan's salmon supply (catch plus imports) totaled 349,000 mt in 1985. This was 38 percent of the world salmon harvest of about 920,000 mt (330,000 mt by the United States, 233,000 mt by Japan, 180,000 mt by the Soviet Union and 177,000 mt from other sources). Major contributors to Japan's supply were: Japan's high seas catch (34,000 mt); imports, mainly from the United States (116,000 mt); and Japan's hatchery returns of chum salmon (168,000 mt). Other sources of supply included 4,000 mt of pink salmon from Japanese hatcheries and 6,000 mt of coho salmon raised in saltwater cages. The hatchery chum salmon runs represented 48 percent of the total Japanese salmon supply in 1985, indicating the vital role being played by Japan's salmon hatchery program. Figure 1 illustrates major changes in sources of supply of salmon to the Japanese market.

Twenty years ago Japan's high seas salmon fishery was the main source of salmon. In 1965 eleven mother-ship fleets with 369 catcher boats and 2,500 land-based vessels harvested 120,000 mt of salmon. About 25,000 mt of this catch was chum salmon. Hatchery chum salmon added only 20,000 mt to the total supply in 1965, and there were almost no

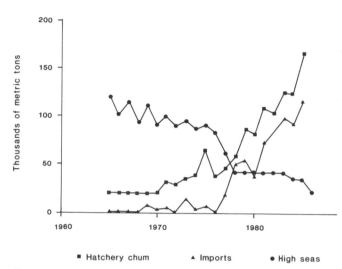

Figure 1. Principal sources of salmon for Japan.

Legend: ■ Hatchery chum ▲ Imports ● High seas

imports. By 1985 the high seas catch had dropped to 34,000 mt (less than one-third of the 1965 high seas catch); but the harvest of hatchery chum salmon had increased to 168,000 mt (eight times the 1965 catch). Furthermore, imports, mainly from the United States, had increased from 7,000 mt in 1975 to 116,000 mt in 1985 (Table 1).

The Soviet Union has established catch quotas (Table 2) and has required Japan to pay fees for harvesting salmon on the high seas since 1978. Because of steady reductions in catch quotas, the size of Japan's high seas salmon fleet has also declined. The high seas catch quota was only 24,500 mt in 1986 with fees of U.S. $21 million paid to the Soviets. Four mother ships (172 catcher boats) and 964 land-based vessels operated in 1986.

Projections indicate that Japan's high seas salmon fleet will be reduced to only three mother ships (129 catcher boats) and 757 land-based vessels in 1987.

Japan's high seas catch is composed largely of chum, sockeye, coho, chinook and pink salmon. The coastal catch is composed largely of hatchery chum salmon.

JAPAN'S SALMON CULTURE PROGRAM

Japan's salmon culture program consists of two types of activities: (1) hatcheries for releasing juveniles into the sea to be harvested as mature adults returning about 4 years later (ranching); and (2) cage culture in salt water where fish are raised to commercial size in captivity (farming). Japan's long-standing salmon hatchery program for ranching will be described, followed by recent developments in farming.

Ranching

Japan's involvement with artificial propagation of salmon for ranching goes back to 1888 when the first national salmon hatchery was built at Chitose, Hokkaido Island. Modeled after the Bucksport Hatchery in the State of Maine, the Chitose hatchery was designed to increase Japan's native stocks of chum salmon. The Chitose hatchery incubated three million eggs in 1888 and six million in 1889 from local stocks. Twenty-eight additional hatcheries had been constructed on Hokkaido by 1892.[1] The first salmon hatchery built on Honshu Island was constructed in 1890.

Table 1. Salmon imports to Japan, 1975-86
(in thousands of metric tons).

Country of Origin	Year										
	1975	1976	1977	1978	1979	1980	1981	1982	1983	1984	1985
United States	4.7	2.4	14.9	40.9	48.1	33.0	60.2	93.3	87.9	80.3	102.5
Canada	1.0	0.3	3.7	7.1	4.7	2.6	5.2	10.8	3.8	5.2	9.9
USSR	0.7	0.4	–	–	0.4	2.0	2.5	0.6	0.3	1.4	1.5
Norway	–	–	–	–	–	–	–	–	0.1	0.3	0.4
China	–	–	–	–	–	–	–	–	0.2	–	0.3
North Korea	0.2	0.5	0.7	1.8	1.4	1.7	3.0	1.5	1.2	1.7	0.8
South Korea	–	–	–	–	–	–	0.8	1.4	1.9	2.0	0.5
Taiwan	0.2	–	–	–	–	–	–	–	3.7	2.4	–
Other	–	–	–	–	–	–	–	0.1	–	–	–
Total	6.8	3.7	19.3	49.8	54.7	39.3	71.8	107.7	99.2	93.2	116.0

Source: Japanese Ministry of Finance.

Table 2. *Japan's high seas salmon quotas and fees to the Soviet Union, 1977-86.*

	1977	1978	1979	1980	1981	1982	1983	1984	1985	1986
Quota (mt)	62,000	42,500	42,500	42,500	42,500	42,500	42,500	40,000	37,600	24,500
Catch (mt)	62,639	41,517	42,447	42,480	42,267	42,368	42,098	35,464	34,318	20,000
Fee to USSR in millions of US $	0	8.3	15.0	16.4	18.2	16.1	17.9	17.9	17.7	21.0

Source: Fisheries Agency of Japan.

Until the end of the Tokugawa Shogun period in 1867, salmon were caught only in rivers. Trapnets were sited along the sea coast in the early part of the Meiji era (1867-1912), and the coastal trapnet catch of salmon first exceeded the river catch in 1882. Total catch of salmon in Honshu increased modestly from 2,700 mt in the 1890s to 3,600 mt in the 1910s, but overfishing caused a progressive decline to about 2,500 mt in the 1920s and 1930s, 1,600 mt in the 1940s and only 950 mt in the 1950s. Poaching of salmon during and shortly after World War II is believed to have contributed to overfishing. Artificial propagation was not expanded during the war.[2]

The Aquatic Resources Conservation Act, which was enacted in 1951 and implemented in 1952, stimulated Japan's modern hatchery program for chum, cherry, and pink salmon. The Act prohibits harvest of salmon in rivers, and instructs the Minister of Agriculture, Forestry and Fisheries to carry out a national salmon hatchery program. The program includes subsidies to private salmon hatcheries and financial contributions from trapnet fishers, who benefit directly from the program. The Act stipulates that juveniles released from hatcheries—whether public or private—and returning mature salmon remain the property of the national government until legally harvested. Accordingly, no one is allowed to catch them without special licenses granted by the minister or governors of prefectures acting for the minister. The Act declares Japan's salmon hatchery program to be the responsibility of the national government under a policy designed to increase fishery resources for the economic improvement of commercial fishers.

Under the Act, the Fisheries Agency of Japan prepared two 5-year programs (1954-58 and 1959-63) for the development of artificial propagation of salmon. Goals included the collection of 800 million chum eggs for Hokkaido hatcheries in 1962 and 200 million for Honshu hatcheries in 1963, subsequently increased to 300 million in 1966. In the same year sockeye salmon were added to the program. Additional plans covering periods of 3 and 5 years have been prepared for subsequent years.

The adoption of extended jurisdictions to 200 miles by many countries in the 1970s has stimulated the Japanese government to further increase the number of juveniles released from hatcheries. Production goals were reevaluated at the end of 1978 and a revised program was initiated in 1979 to increase domestic salmon stocks in order to counteract restrictions on high seas fisheries.[3]

Releases of salmon from government and private hatcheries had reached 400-500 million juveniles in the 1960s, but their numbers increased to about one billion by 1977 and two billion since 1983. In 1986, there were 37 national governments, six prefectural governments, and 105 private (fisheries cooperatives) hatcheries on Hokkaido and 165 private hatcheries on Honshu. Chum salmon dominate production. Numbers of juveniles released since 1952 are shown in Figure 2.

In 1979, the proportion of the budget of the Fisheries Agency appropriated to the salmon hatchery program was increased from 0.7 percent to 1

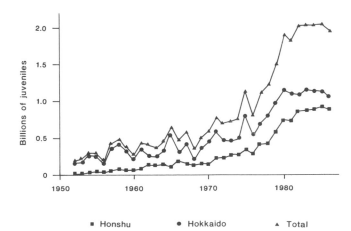

Figure 2. *Numbers of juvenile chum salmon released from hatcheries in Japan, 1952-1986.*

percent. In 1986 the budget for the salmon hatchery program was U.S. $17.1 million, about five times the budget in 1976 (U.S. $3.5 million).

The Aquatic Resources Conservation Act empowers the national government to give subsidies to private fisheries cooperatives for building and operating salmon hatcheries. In 1986, U.S. $6.5 million (38 percent of the hatchery program budget) was given as subsidies to fisheries cooperatives operating 270 private hatcheries on Honshu and Hokkaido. Mature chum salmon now return to hatcheries on Hokkaido and sixteen prefectures on Honshu.

Chum runs from the salmon hatchery program in 1985 totaled 49.3 million fish weighing 168,000 mt. The runs have trended upward (Figure 3) in response to increased numbers of juveniles released from hatcheries and improved survival. Average return rate for hatchery chum salmon has increased from 1.80 percent in 1970 to 2.77 percent in 1980. The most recent available return rate (1985) was 2.71 percent (Hokkaido 2.95 percent and Honshu 2.36 percent) (Figure 4). This highly successful return rate results from improved hatchery technology and timing of release to correspond with favorable environmental conditions at sea.

Fry are fed an artificial diet consisting primarily of fish meal for 1 to 2 months before being released. Short-term rearing of fry has made it possible to produce healthy fingerlings of a size which favors survival at sea.

Chum salmon returning to Hokkaido are divided into three seasonal groups: early (September), middle (October-November), and late (after November). By including all three groups in hatchery production, it has been possible to provide trapnet fishers and consumers with a stable supply of chum salmon over an extended period.

Skin color of hatchery chum salmon tends to remain silver bright (which consumers prefer) until the end of October, when the color changes to semibright or dark. The National Salmon Hatchery has since 1981 emphasized the collection of eggs from early runs, as salmon from eggs collected in the early season tend to return early.

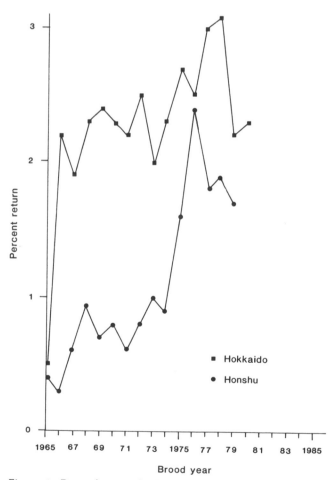

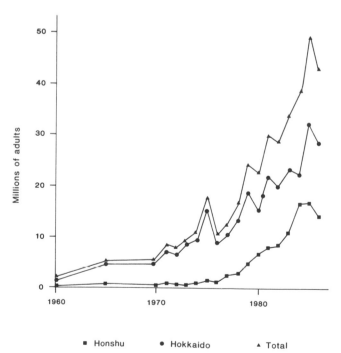

Figure 3. Hatchery chum salmon harvested in Japan.
Data from Fisheries Agency of Japan.

Figure 4. Rate of return for Japanese hatchery chum salmon.
Data from Fisheries Agency of Japan.

Hatchery pink salmon totaled 2.8 million adults weighing 4,200 mt in 1985. Pink salmon fry have been released only from hatcheries on Hokkaido. Return rates have compared favorably with chum salmon. Hatchery cherry salmon numbered only 5,000 fish in 1985 from 13,438,000 juveniles released in 1982. The return rate was a disappointing 0.04 percent. Sockeye salmon have been released from hatcheries in limited numbers since 1969. Returning runs of adults have remained small.

Some natural spawning occurs in Japanese salmon streams, particularly by pink and cherry salmon. Chum salmon are considered to be of hatchery origin, even though some incidental natural production probably occurs.

Domestic interception of chum salmon returning to several hatcheries is a problem. With a large number of chum salmon returning to Japan, governors of Hokkaido and northern Honshu prefectures have extended fishing privileges beyond the traditional trapnet operations by granting permits to coastal longliners and poleliners. Trapnet fishers in the southern part of Honshu now claim that these newly authorized operations intercept salmon destined for their hatcheries.

Funds for the operational costs of hatcheries come primarily from national and prefectural governments and fisheries cooperatives. The percentage of total costs paid by cooperatives has increased from 35 percent in 1981 to 46 percent in 1985.

A Fall Chum[4] Salmon Resource Management and Coordination Council was formed in April 1986. It consists of two associations (Hokkaido and Honshu) representing fisheries cooperatives involved in the salmon hatchery program, one trapnet fishers' association, and one association representing other coastal fishers. National and prefectural officials and scientists sit as ex officio observers. The council has two (Pacific and Japan Sea) regional councils to address problems related to interception and finances and to coordinate research.

Areas of research requiring emphasis include production of healthier fry; determination of ideal release times to maximize ocean survival; selection of brood stock to increase production of silver bright adults; increasing returns to the Japan Sea coast, where survival is lower than on the Pacific coast; determination of migration routes to various hatchery streams; and improved resource management for conservation.

Due to budget constraints, the National Salmon Hatchery plan for production of chum salmon in Hokkaido has remained at 1,049.3 million fry since 1984. However, actual releases have somewhat exceeded this target (Figure 2). Should the national government's share of financial support decline in relation to that of trapnet fishers and cooperatives, the government will still remain an active participant. With a national government policy of increasing domestic fishery resources within Japan's 200-mile zone, Japan's salmon hatchery program is expected to continue to increase from the 1985 catch of 52.1 million fish (all species combined).

Cage farming

The development of farming in Japan is modeled after Norwegian success in pen-rearing Atlantic salmon. It focuses mainly on coho salmon but also involves sockeye and chinook.

Nichiro Fisheries Company of Tokyo has pioneered the farming of salmon in Japan. Freshwater culture of sockeye,[5] chinook, chum, and pink salmon was started in 1971 after the company's high seas salmon mother-ship fleet was reduced due to imposition of quotas. In 1973, Nichiro began to concentrate on coho salmon farming because of its superior disease resistance. Nichiro initially imported eyed coho salmon eggs from U.S. public salmon hatcheries in Washington and Oregon and raised them to smolt size in a freshwater lake in Shizuoka Prefecture. Nichiro has transported coho salmon fry reared in freshwater facilities to cages in the sea off Shizugawa,[6] Miyagi Prefecture (northern Honshu). The most common size of pens is 10 m x 10 m x 7 m deep.

Acclimatization to seawater is reportedly achieved by placing the fry in 50 percent seawater on the first day, raising the salinity to 70 percent on the second day and by 10 percent a day until full salinity is achieved on the fifth day.[7]

A marketable fish is produced in 19 to 20 months from hatching. A typical production cycle is outlined below:[8]

December	Eyed eggs to hatchery
January	Hatching
Till October	Rearing in freshwater until fry grow to 100-200 g
November	Transfer to seawater pens

April	Growth to 1.3 kg
May	Growth to 1.4 kg (marketing starts)
June	Growth to 1.9 kg
July	Growth to 2.5 kg
August	Growth to 3.0-3.5 kg (marketing ends)

Nichiro reportedly produced 450 mt of coho salmon in 1978 and its yield from fry to market size was 80 percent. In 1986, fish production cost was reportedly U.S. $2.09/lb, while the market price for fresh pen-reared coho salmon ranged from a high of U.S. $3.27/lb in May to a low of U.S. $1.67/lb in August. In order to circumvent low prices in summer, about 2,000 mt of the total 1986 production was reportedly salted and is expected to sell at U.S. $3.27/lb.[9]

Nichiro's venture was followed in 1979 by Taiyo Fishery Company,[10] also a high seas salmon fishing firm, and in 1984 by Nichimo Company, a manufacturer/supplier of nets and other fishing gear, and the Federation of Miyagi Fisheries Cooperatives.

The number of farmers of coho salmon has increased from 150 in 1983 to 180 in 1984 and to 224 in 1985.[11] Sites for cages are primarily off the coast of Miyagi and Iwate prefectures. Production has increased one-hundred fold from 72 mt in 1978 to 7,200 mt in 1986.

A fisheries newspaper[12] has predicted that cage culture of coho salmon will not increase beyond 15,000-20,000 mt/year in Japan due to the limited area available with suitable temperatures (in the range of 5-10°C) and the limited supply of eyed coho eggs from the U.S. or other sources. Major efforts to develop coho brood stock in Japan have not been made but are a possibility.

Nichiro has also enjoyed some success with the farming of both sockeye and chinook salmon. In January 1983, the company collected eggs from landlocked sockeye (kokanee) in Lake Chuzenji, Nikko, Tochigi Prefecture. In November 1983, 2,500 juveniles, each weighing 80 g, were transferred to cages in Shizugawa Bay, Miyagi Prefecture. By August 1984, these juveniles had grown to an average of 500 g, with some fish weighing over 1 kg.[13] Nichiro sources say that much care was needed during transportation since sockeye juveniles are vulnerable to injury from rubbing one another or the walls of a container and during acclimatization to seawater. Also, a special food was formulated to satisfy the nutritional needs of a species which feeds primarily on zooplankton. A source at the company thought that it would become feasible to produce farmed sockeye weighing 2 kg or more.

Nichiro's experience with chinook dates back to December 1983, when 10,000 eyed eggs were shipped from Seattle, Washington, and reared in fresh water for 2 years in Shizuoka, near Mt. Fuji. During this period they grew to 350 g. In November 1985, 10,000 juveniles were shipped to two saltwater sites, Sado Island in the Japan Sea, Niigata Prefecture, and Shizugawa Bay in the Pacific Ocean, Miyagi Prefecture. By June 1986, the survivors had grown to 2.6 kg with the largest one weighing 6.0 kg. Survival in cages was 80 percent. Moist pellets consisting of a mixture of sardine, formula feed, and krill-like mysis were fed. The fish were marketed fresh in June 1986 and brought prices ranging from U.S. $3.22 to U.S. $4.03/lb. Nichiro plans to devote a longer period of time to the freshwater rearing stage. There is also interest in farming chinook salmon in Japan, since the supply of chinook salmon to Japan from other sources is very limited (totaling about 2,500 mt). Nichiro hopes to produce 20,000 chinook salmon in 1989, and a successful effort on their part could lead to significant future expansion of farming.

CONCLUSIONS

Returns of adult hatchery salmon to Japan's coastal fisheries are expected to expand beyond the 52.1 million salmon harvested in 1985. The program appears destined to grow regardless of how financing is structured among national and prefectural governments and the private sector. Emphasis will be placed on improvement of quality and larger size.

The growth of farming is expected to be constrained by lack of coastal waters suitable for cage culture of salmon and limitation in the supply of eyed eggs. In addition, the government has offered no financial incentives for the development of salmon farming, by contrast with its policy on ranching. National policy appears to be to let the market decide. Instead of encouraging a rapid increase in salmon farming, the Fisheries Agency appears to be more concerned about diseases coming into Japan with importation of eyed eggs.

Japan's high seas salmon fishing is faced with the prospect of further curtailment by agreement with the U.S.S.R., U.S., and Canada. There is an uncertainty about Japan's continued access to salmon on the high seas. The continued development of domestic ranching and farming industries is clearly in Japan's interest. Availability of water, space, and brood stock are likely to be the factors that limit future growth of salmon aquaculture in Japan.

ACKNOWLEDGMENTS

I would like to extend my appreciation to Mr. Kazutoshi Nara (Salmon Management Section, Fisheries Promotion Division, Fisheries Agency of Japan) for providing me with data on Japan's salmon hatchery programs. My thanks are also to Dr. John G. Gissberg, Regional Fisheries Attache, American Embassy, Tokyo for his advice and editing.

LITERATURE CITED

1. *Sake Masu Zoshoku no Ayumi* (1969), the Japan Salmon Resource Conservation Association.
2. *Sake Masu Shigen to Sono Gyogyo* (1966), Kisaburo Taguchi.
3. *Sake Masu Zoshoku Jigyo no Tenkai Hoko* (Future Development of Artificial Propagation of Salmon) (January 1985), Fisheries Agency of Japan.
4. *Suisan Shuho*, No. 1077, 7/25/86; *Suisan Kai*, No. 1219, August 1986.
5. *Suisan Shuho*, No. 867, August 1979; *Asahi Shimbun*, 5/7/75; *Suisan Tsushin*, 11/8/76.
6. *Suisan Keizai Shimbun*, 1/28/78.
7. *Suisan Shuho*, No. 867, August 1979.
8. *North American Seafood Information*, 4/28/86.
9. *Suisan Keizai Shimbun*, 8/5/86; *North American Seafood Information*, 9/29/86.
10. *Suisan Keizai Shimbun*, 4/16/84.
11. *Suisan Keizai Shimbun*, 4/18/85.
12. *North American Seafood Information*, 4/28/86.
13. *Suisan Keizai Shimbun*, 8/14/84.

Salmon Farming in Norway
Present Status and Outlook into the 1990s

Frank Gjerset

Business Development Manager
Atlantic Salmon Company
Trondheim, Norway

I visited North America in 1984 to familiarize myself with progress on aquaculture. My travels extended from Alaska to Mississippi. My impressions of salmon and catfish aquaculture in North America have altered my perspective on Norwegian aquaculture considerably. My purpose in this paper is to give you some personal opinions on events in Norway leading up to the present state of development of salmon farming. I will develop scenarios on possible future directions of salmon farming in Norway and changes in the structure of the Norwegian industry that are likely to be mandated by competition for markets.

Norwegian salmon farming has been a great success story. The industry has generated good profits, and products have been widely accepted in the marketplace. Production has increased tenfold from about 4,000 mt in 1970 to about 40,000 mt in 1986 (Figure 1).

Various projections have been made about future trends in production out to 1990. They range from about 80,000 to 120,000 mt by 1990, with an average of 100,000 mt (Figure 2). The industry collected enough eggs in autumn 1986 to produce 90-100,000 mt in 1990. If achieved, this would be about 50 percent more salmon than the present

45,000 and 60,000 mt in recent years. It appears that Norwegian production of farmed Atlantic salmon will soon surpass the combined catch of coho and chinook in traditional fisheries (Figure 3).

It is difficult to analyze the success of salmon farming in Norway without having a good understanding of how the rural economy in Norway works and how the traditional fishing industry acts as a driving force to the economy. Aquaculture has little influence on traditional fishing, but the existing infrastructure of the fishing industry is very important to development of fish farming in Norway. Norwegian aquaculture utilizes a technology which is inexpensive and adaptable to small-scale, "backyard" operations. It has enjoyed access to a support system that provides a competitive advantage over most other new industries. Elements of this support system include:

— A hatchery technology that was easily commercialized.

— Access to wild stocks of Atlantic salmon for initial brood stock.

— A strong professional interest in selective breeding by government and university laboratories and private growers; wild fish are no longer needed for brood stock and have been replaced by farmed stocks with desirable genetic traits.

— Little regulatory interference by government during early development of the industry; innovation was encouraged and numerous technological advances resulted.

— A reliable supply of low cost feed in the form of fish and fish by-products from the traditional fishing industry.

— Existing distribution networks for supplying Norwegian fish products to European markets which easily accommodated expanded production of farmed Atlantic salmon.

— Ready acceptance of farmed salmon by the market which had been constrained more by supply than demand.

— Recognition of the economic potential of salmon farming by the Norwegian financial community in the early stages of development of the industry.

It is common practice in Norway for government to protect domestic industries against imports, to provide subsidies, and to regulate production. Strong governmental regulations apply to

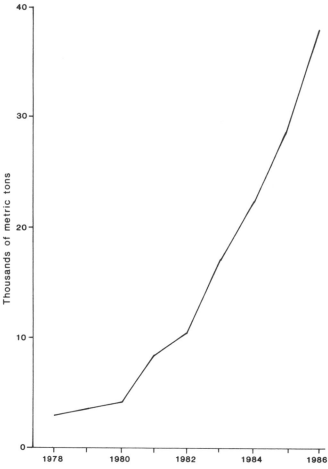

Figure 1. Production of farmed Atlantic salmon in Norway, 1978-86.

catch in Canadian west coast waters. Norway is already producing enough farmed salmon to virtually satisfy the demand for fresh and frozen salmon in European markets. Furthermore, exports to North America are rapidly increasing. Thus, farmed Atlantic salmon from Norway and other European countries, when added to the modest harvest of wild Atlantic salmon in Europe, are displacing imports into Europe from North America and are being exported to the U.S. in increasing quantities. Other European countries which produce farmed salmon include Scotland, which ranks a distant second behind Norway and produces less than 10,000 mt, and the Faroes, Iceland, and Ireland, which produce lesser amounts.

Norwegian salmon compete in many world markets primarily with coho and chinook. The world harvest of coho and chinook has varied between

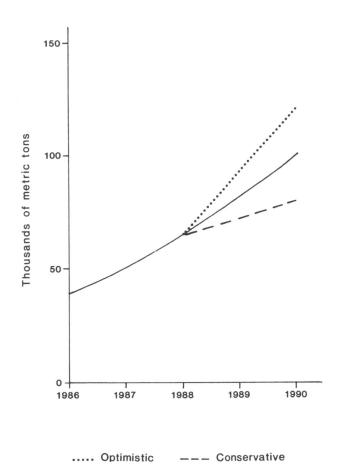

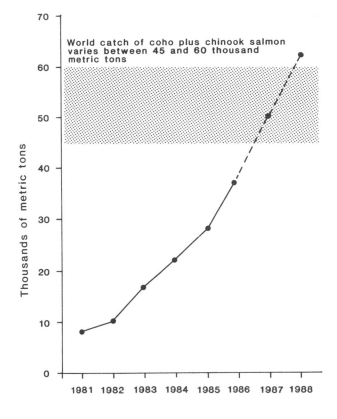

..... Optimistic   — — — Conservative

Figure 2. Projected production of farmed Atlantic salmon in Norway through 1990.

● Norwegian production of farmed Atlantic salmon

Figure 3. Norwegian production of farmed Atlantic salmon is expected to surpass world catch of coho and chinook salmon by 1988.

agriculture, forestry, and traditional fisheries and there is a similar tendency to regulate the salmon farming industry. However, since 98 percent of farmed Norwegian salmon is exported, it is essential that the regulatory environment allow the product to compete effectively in world markets. Norwegian salmon farming has therefore enjoyed relative freedom from government regulation. Fifteen years of development have produced favorable economic results. However, few economies of scale have been realized due to limitations on size of individual farms. One domestic regulation places a limit of 8,000 m³ on the size of individual salmon farming businesses in order to avoid concentrating ownership of salmon farms under a few large corporations and to insure a broad distribution of ownership in rural communities. More will be said about this later.

Prices received by Norwegian salmon growers are projected to decline in coming years due to increased world supplies of farmed salmon. Norwegian growers will be forced to become more efficient in order to remain competitive. Atlantic salmon are expensive to grow. Three factors contributing to high production costs are feed, smolts, and finance charges. Of these, the cost of smolts is the most likely candidate for reduction. The present high cost of smolts will need to be reduced by economies of scale if Norwegian salmon farming is to remain competitive into the 1990s.

Opportunities to reduce mortality of farmed salmon through control of disease and other improvements in technology also need to be explored more intensively. Disease has not been a widespread problem in Norway, but individual farms have experienced difficulties to varying degrees.

Norwegian salmon farmers receive some help with disease control from public agencies, and this is augmented by disease specialists working for private industry. Also, salmon farmers are gaining experience with disease control methods. Certain disease problems have been relieved through development of improved fish feeds and culture methods.

Specific diseases of continuing concern include the following:

— Infectious pancreatic necrosis (IPN) occurs at a number of hatcheries. Atlantic salmon appear to be somewhat resistant to this disease, and it has not been a major cause of mortality. It nevertheless represents a continuing threat.
— Bacterial kidney disease has contributed to temporary closure of hatcheries and farms.
— Furunculosis contributed to the closure of twenty-nine farms in 1986, and substantial costs have been attributed to this disease.
— Red mouth disease has been a problem with imported smolts.
— Various types of parasites create problems, but most can be controlled by prophylactic techniques. There are indications that wild stocks have become infected with a parasite (gyrodactolus), possibly originating from hatcheries.
— Vibriosis is a major problem in salt water. Vaccines have not provided sufficient protection to warrant a high level of confidence in their use.

There has been good progress with development of high quality salmon diets in Norway. The fishing industry generates large volumes of carcass by-products which are used in salmon diets and capelin, which are abundant in Norwegian waters, are harvested to feed salmon. Both sources of fish protein are commonly used in formulations for dry and moist diets and for silage. Manufacturers of salmon feeds have placed considerable emphasis on high quality dry diets. Norwegian hatcheries feed predominantly dry diets, whereas hatcheries in the western U.S. also feed moist diets. Dry diets account for more than 85 percent of Norwegian fish food production, and their portion of the market continues to grow. An important advantage of dry diets is that they are easier and cheaper to store and handle. The price of a high quality dry diet is about U.S. $0.90 per kg. Approximately 1.3 to 1.4 kg of dry feed is required to produce 1 kg of salmon.

Research on selective breeding by government and industry has also contributed substantially to progress of Norwegian salmon farming. As many as five generations of selectively bred strains of Atlantic salmon have been produced. Traits which have been selected for include time of maturation, food conversion rate, growth rate, and disease resistance. Improvements in growth rate have been most impressive. It formerly took 24 months to grow a smolt to 3 kg for market in seawater. This can now be achieved in less than 16 months. Some believe that a 3 kg fish can be produced in 12 months or less.

Smolt production is a serious impediment to progress. The supply of smolts is growing, but it is still limited. Twenty-two million were produced in 1985, compared to seventeen million produced in 1984. Prices are high and out of line with other costs. Survival from egg to smolt has not been very satisfactory. An egg-to-smolt survival of only 30 percent is typical. For example, only seventeen million smolts were produced in 1984 from approximately sixty million eggs.

In order to maintain a competitive position in a world salmon aquaculture industry, Norway must increase smolt production and reduce costs. Prices have been as high as U.S. $4.00. Because of high prices due to scarcity, smolt production has been profitable. There is currently much speculative investment in new hatcheries and scarcity could quickly change to surplus. If new hatcheries in planning and under construction achieve full production, Norway will produce enough smolts to sustain an annual production of at least 200,000 mt of harvestable salmon. The price of smolts could decline precipitously to a point where cost of production could exceed market values for many inefficient hatcheries. Some hatcheries are unlikely to compete successfully if the price declines below U.S. $2.00 per smolt. A price of about U.S. $1.00 per smolt should allow the Norwegian farming industry to compete successfully in future world markets.

Because of the lengthy production and marketing cycle, there is a significant time lag between market pressures for lower prices and the response of producers to cut costs, including cost of smolts. In 1986 the average market price of harvested Atlantic salmon declined by 30 percent, while the price of smolts reached new highs. It will take time for fish farmers to realize that profits are disappearing. Smolts purchased today will be ready for the market in 1988. Today farmers receive around $5.70 per kg for a high quality, gutted fish (head on). But by 1988 they will be lucky to receive U.S. $1.80/kg. Because it takes 3 to 5 years before a new cycle of production from egg to harvestable fish is completed, investments in production already initiated cannot be reversed. Investors will need to exercise considerable caution with new commitments to produce smolts at high cost or to buy them at high prices.

Shortage of smolts has been a chronic problem for Norwegian salmon farmers. However, recent investments in new and expanded hatcheries along with declining prices for farmed salmon should soon contribute to an oversupply of smolts. The developing overcapacity for smolt production is partly the result of poor government policies which have stimulated overly optimistic expectations within many rural communities. Before 1977, anyone could start a hatchery or fish farm. In 1977 a law was introduced to limit the number of new entrants because of early fears of oversupply. However, the market has grown dramatically and has met or exceeded all projections. Approximately 350 new farms have been added to the approximately 400 registered before the 1977 law came into effect. Only in the last few years has it become difficult to obtain permits for new farms, and there are continuing political pressures to spread ownership. Such a social policy contributes to inefficiencies in the industry.

There is an apparent lack of understanding of the dynamics of the development and maturation of a new industry and the necessity for consolidation if new markets and growth are to be nurtured. There is a tendency to preserve a structure that is socially oriented but inefficient. The risk of such a policy is that other countries are likely to outcompete Norway in future world markets including North America, which has the potential to become the most important market for farmed salmon. Norway's future as a salmon farming country should be based on cost effective production and competitive pricing.

The U.S. catfish industry is a useful model for developing scenarios on future development of salmon farming in Norway. Catfish and salmon are exposed to similar forces; both industries have undergone or will undergo similar changes; neither industry has matured fully. But there is one significant difference: Norwegian farmed salmon have supplied a latent market demand, where it has been necessary to create a new market for catfish. The leading U.S. processor of catfish is currently marketing about the same poundage as all Norwegian salmon farming producers combined. Characteristics of Norwegian salmon and American catfish farming are compared in Table 1.

Some industry and government institutions believe that the Norwegian salmon farming industry can continue to grow, prosper, and compete successfully in mass markets without altering its structure. This is unlikely, in my opinion. I believe that the Norwegian salmon industry will need to follow the example of the American catfish industry, which was restructured in the early 1980s. Restructuring triggered the present mass marketing of catfish in the U.S. Similarly, if the Norwegian salmon industry is to grow successfully, avoid declining profits,

Table 1. Characteristics of Norwegian salmon and American catfish farming.

Salmon farming (Norway)	Catfish farming (U.S.)
A commodity.	Branded products.
No emphasis on reducing production costs through economies of scale.	Great emphasis on reducing production costs through economies of scale.
Satisfying existing market demand for high cost fresh salmon.	Creating new mass markets for low cost fresh fish.
Little processing required.	Processed into fillets and portions.
Traditional fresh fish markets already established.	New markets for fresh and frozen fish products are promoted.
Processing and marketing is fragmented.	Processing and marketing is highly concentrated.
Industry is supply driven.	Industry is market driven.
Markets likely to be satiated within 3 years if existing structure is retained.	Huge new markets being developed through innovative promotion and competitive pricing.

and achieve production levels much above 70,000 mt, production, processing, and marketing will have to undergo major restructuring within 3 years.

Production of farmed Norwegian salmon increased from 28,000 mt in 1985 to about 40,000 mt in 1986. More than 50,000 mt is expected in 1987. The European market for fresh and frozen salmon is approximately 45,000 mt. In 1985 approximately 17,000 mt of farmed Norwegian salmon were sold in Europe, and 8,000 mt in the U.S. An additional 3,000 mt were sold elsewhere. Scotland produces about 10,000 mt, primarily for the European market; and the commercial harvest of wild Atlantic salmon of around 10,000 mt annually is also sold primarily in Europe. Atlantic salmon are displacing frozen Pacific salmon from European markets, and this substitution within existing markets is expected to increase. Eventually, farmed Atlantic salmon are expected to displace north American sockeye, coho, and chinook in European markets and may even become a substitute for the cheaper pink and chum salmon. However, the European market is unlikely to absorb expected increased production of Norwegian farmed salmon even if displacement of Pacific salmon continues. New markets will need to be developed. Growth of the U.S. market has been very welcome and has relieved pressure to develop additional marketing opportunities in Europe. But there are now many signs that Norway must greatly enlarge its markets if expanded production expected in coming years is to be sold at a profit.

The Norwegian salmon industry could face some very hard times in the next 2 to 4 years. More than seventy companies have an export license, and internal industry competition for markets is severe. Export companies are already experiencing difficulties with profits. I expect that at least 50 percent of farmers will experience little or no profit or possibly a loss in 1988. This is partly the result of government policy, as noted above, and partly the result of failure of the industry to anticipate change in markets. Norwegians have traditionally been suppliers of raw materials and marketing opportunities have not been developed. Development will require production and marketing strategies that the traditional Norwegian fishing industry has never used.

Americans, on the other hand, have been very successful in development of mass markets. Poultry is an example. In the 1950s chicken was a luxury food in the U.S.; today it is a low-priced staple. Production costs have been reduced through application of technology, and mass markets have been developed. The same transition is expected with marketing salmon in the U.S. and Europe. Existing complex marketing systems involving many small competitors will need to be replaced by an efficient distribution and marketing system involving relatively few participants. American catfish and trout farming industries have already gone through this transition.

Norway can have a great future as a mass marketer of salmon if proper production and marketing strategies are implemented. It will be more difficult for Norway to compete in North American markets than in European markets because of higher transportation costs.

Many factors inhibit consolidation of the Norwegian industry, reduction of production costs, and development of mass markets. Fortunately, Norway has not had to deal with serious conflicts over resource development and allocation which seem to present major obstacles to salmon aquaculture in North America where there has seemingly been much debate about ecological and genetic questions. Salmon farming in Norway has not had to deal with problems of this sort.

Five alternative scenarios concerning the future of salmon farming in Norway are presented below, two for the near-term up to 1990 and three for the long-term after 1990:

1. The industry peaks at between 60,000-80,000 mt annual production by 1990. There is little or no reduction in cost of production. Production capacity fails to expand for lack of profitability.
2. Market prices of salmon reach such low levels by 1990 that the industry becomes unprofitable, but production capacity continues to increase until bankruptcies force a halt to growth. Government finally alters policies to encourage consolidation of the industry. Production costs decline and the industry begins to grow again after profitability is attained.
3. No further growth occurs after 1990 because government policy has discouraged consolidation and cost reductions at all levels. No new markets are developed.

4. Production is consolidated after 1990 and costs are reduced and stabilized. World prices of farmed salmon decline to about U.S. $3.00/kg. or possibly less. There are fewer but larger farms; many small and marginal producers have been forced out of production. Marketing remains fragmented with little development of mass marketing of fresh and processed products. The industry matures with limited potential for growth.

5. Production and marketing becomes concentrated after 1990 in vertically integrated organizations, each controlling quantities of at least 10,000 mt. There may be five to seven such organizations. This consolidation becomes a driving force for profitable mass marketing. Production costs are reduced significantly. Processing for value added products becomes increasingly important. Production continues to grow at least moderately in response to expanding markets.

It is my opinion that scenario number 2 is most likely before 1990. In the long-term, however, it is difficult to know which direction the industry will move. Only scenario 5 will pave the way for developing the full potential of salmon farming in Norway and insuring that Norway maintains its market lead over other nations. Salmon farming could become a U.S. $1 billion industry in Norway within 12 years, producing 250,000 mt a year. This will happen only if government and industry develop and implement policies which encourage growth. The American catfish industry is a useful model here for Norwegian planners and policy makers.

It is my recommendation that laws and administrative rules restricting consolidation of the Norwegian salmon farming industry be modified to encourage industrialization of the industry. Political concern over maintenance of small farms and other social issues could prevent this. If so, Norwegian investors will continue to explore investment opportunities in salmon farming projects in foreign countries such as Chile, the U.S., and Canada instead of in Norway.

Recent Changes in the Pattern of Catch of North American Salmonids by the Japanese High Seas Salmon Fisheries

Colin K. Harris

Fisheries Research Biologist
Fisheries Research Institute
University of Washington

INTRODUCTION

In 1952 Japan established two large high seas drift gillnet fisheries for Pacific salmon, a mothership fishery that has operated north of 46°N and west of 175°W, and a land-based fishery that has operated mostly south of 46°N and west of 175°W. Regulation of these fisheries has been imposed in part by two international treaties, the International Convention for the High Seas Fisheries of the North Pacific Ocean (North Pacific Treaty), signed between Canada, Japan, and the United States in 1952, and the U.S.S.R.-Japan Treaty Concerning the High Seas Fisheries of the Northwest Pacific Ocean (U.S.S.R.-Japan Treaty), signed in 1956. The North Pacific Treaty regulated the high seas salmon fisheries for many years by setting their eastern boundary at about 175°W, but negotiations and frequently changing protocols under the U.S.S.R.-Japan Treaty have led to numerous and diverse regulations, including catch quotas, time/area closures, gear restrictions, and others. While these treaties have been broad in scope, the Japanese high seas salmon fisheries have been the principal subject of concern in both forums.

Consistent with an international trend in jurisdiction of marine resources, the United States,

Canada, and the U.S.S.R. established 200-mile fishery zones around their coasts in 1977-78, and claimed a greater degree of ownership and jurisdiction of anadromous salmonids spawned in their respective territories. These changes in fisheries policy prompted revision of the North Pacific Treaty and the U.S.S.R.-Japan Treaty, and caused in turn a reduction of the overall magnitude of the Japanese high seas salmon fisheries and a major change in the time/area pattern of fishing. Since 1977-78, the United States and the U.S.S.R. have pressed Japan for further reduction of the fisheries, and agreements since 1978 have led to further measures to reduce high seas salmon catch.

This paper provides a brief history of the Japanese high seas salmon fisheries, a review of the treaty changes and a description of how they have affected the fisheries and the catches, and a preliminary, mostly qualitative analysis of how the level of high seas catch of salmonids of North American origin has changed. The period 1972-85 is chosen for primary focus in this paper because statistics for the land-based fishery have been presented in a standard format and level of detail since 1972, and the period provides adequate data for comparing trends before and after the treaty changes of 1977-78. All fishery statistics summarized are from International North Pacific Fisheries Commission (INPFC) (1979), INPFC Statistical Yearbooks, INPFC Documents, or from Japan Fisheries Agency (personal communication), unless otherwise noted.

HISTORY AND DESCRIPTION OF THE JAPANESE HIGH SEAS SALMON FISHERIES

The Japanese mother-ship and land-based driftnet fisheries are a recent chapter in a long history of Japanese distant salmon fishing. Japanese salmon fishing on the east coast of Siberia began as early as the 17th century, and Japanese salmon fisheries on Russian territory were well established and made large catches in the late 1800s and early 1900s (Wertheim 1935, Gregory and Barnes 1939, Leonard 1944, Atkinson 1964, Parker 1974). Tension with Russian nationals in the 1920s and 1930s made the Japanese shore-based operations on U.S.S.R. territory increasingly difficult; this in part prompted two new fisheries, a mother-ship driftnet fishery off the Kamchatka Peninsula in the U.S.S.R. and a driftnet and trap fishery in the northern Kuril Island

passes, that harvested salmon at sea during their spawning migration. These large fisheries were short lived, however, as the outbreak of war in the Pacific interrupted all Japanese distant salmon fishing.

After the war, Japan began to develop distant water fisheries, including salmon fisheries. In 1952 three new salmon fisheries—a mother-ship driftnet fishery, a land-based driftnet fishery, and a land-based longline fishery—received licenses and began operations in the north Pacific Ocean. The longline fishery is not mentioned further in this paper as it ceased to exist after 1971.

Detailed decriptions of the mother-ship fishery can be found in Fukuhara (1955, 1971), Neo (1964), Manzer et al. (1965), Nagasaki (1967), Chitwood (1969), and Fredin et al. (1977). Although the fishery has changed in a number of ways in its 35-year history, it has always consisted of large mothership processors each accompanied by a number (32-34 in 1972-76, 43 in 1978-86) of smaller catcher boats. Each catcher boat is assigned to a particular mothership for the entire season, so the fishery operates in discrete fleets. Some of the catcher boats (six in each fleet in recent years) operate as scout boats and can range far from the rest of the fleet, but the regular catcher boats fish in an array near the mother ship and deliver their catches each day. The fishing gear has changed much over the years, but in recent years it has consisted of monofilament nylon with (stretched measure) mesh sizes of at least 130 mm in at least 60 percent of the net and at least 120 mm in the remainder. The nets hang about 8 m deep in the water. The nets are in sections, called tans, about 47 m long, and a catcher boat's set consists of 330 tans tied end-to-end for a total net length of about 15 km. Nets are set in late afternoon or early evening, soak overnight, and are retrieved in the early morning. After catches are delivered to the mother ship, the catcher boats proceed to the position assigned by the fleet commander for the next evening's set. The mother ships primarily canned salmon in the early years, but since 1978 they have prepared exclusively frozen product. The fishery area is divided into regulatory blocks, and the utilization of the blocks is controlled to prevent overlap of the fishing arrays occupied by catcher boats from different fleets. The selection of blocks to be fished is decided jointly by fleet commanders and officials from the Japan Fisheries Agency. Times and areas of operation are described later.

The mother-ship fishery developed rapidly, from three mother ships and 57 catcher boats in 1952 to twelve fleets with 406 catcher boats in 1955 (INPFC 1979). The fishery reached its maximum size in 1959 (sixteen fleets), and in 1960 began a gradual decline to ten fleets in 1972-76, largely in response to quota reductions imposed through the U.S.S.R.-Japan Treaty. The fishery's catches dramatically increased in the initial years, from 2.1 million salmon in 1952 to 64.0 million fish in 1955 (INPFC 1979). Catches averaged 53.7 million salmon in the peak 1955-59 period, and declined to a level of 18.9 to 26.4 million fish in 1960-76 (Figure 1). The total catch declined to 16.8 million fish in 1977, and further declined to between 6.9 and 9.6 million salmon in 1978-85.

As its name implies, the land-based driftnet fishery consists of vessels that land their catches at Japanese ports rather than delivering them to floating processors (Fredin et al. 1977). Unlike the

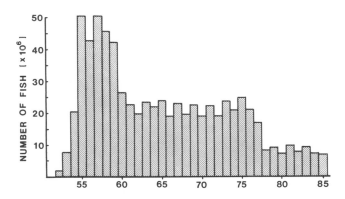

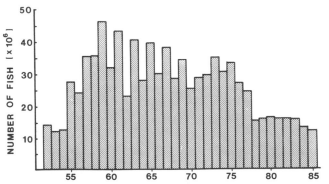

Figure 1. Catches, in millions of fish, of salmon by the Japanese mother-ship salmon fishery (upper chart) and the land-based driftnet salmon fishery (lower chart), 1952-85.
Sources: Statistics for the land-based fishery for 1952-69 are from Fredin et al. (1977); all other statistics are from International North Pacific Fisheries Commission (1979) or INPFC Statistical Yearbooks or Documents.

mother-ship fishery, the land-based vessels operate independently. There are actually two distinct land-based driftnet fisheries, one consisting of small (under 10 ton) prefecturally licensed vessels that operate west of 153°E (147°E before 1978), and larger (over 30 ton) federally licensed vessels that may operate much farther eastward. The small-vessel fishery will not be mentioned further in this paper. The gear used by the large-vessel fishery is similar to that used by the mother-ship fishery, although the mesh size is smaller (at least 110 mm). The maximum length of a net is 15 km. Nets used by the land-based fishery are also composed of sections called tans, but the lengths of tans vary from 30 m to 45 m (Japan Fisheries Agency, personal communication).

The large-vessel land-based driftnet fishery also developed quickly, from 58 vessels in 1952 to over 200 by 1954 (Taguchi 1966, Fredin et al. 1977). Fleet size was at its maximum in 1972-74 (374 vessels), but fishing effort (in terms of vessel-days) was highest in the mid-1960s. Statistics for the early fishery are at various levels of detail and in various formats, and therefore are difficult to compare with later statistics. Fredin et al. (1977) estimated the early catches (by species) by the large-vessel fishery, and showed that the catches increased from 12.2 to 14.3 million salmon in 1952-54 to levels of 23.2 to 46.3 million fish in 1955-76 (see Figure 1). The 1977 catch was 24.4 million, and the 1978-85 catches ranged between 12.4 and 16.5 million salmon.

INTERNATIONAL TREATIES AFFECTING JAPANESE HIGH SEAS SALMON FISHERIES
North Pacific Treaty

As Japan began to develop distant water fisheries after World War II, the United States realized the potential for conflict with U.S. fishing interests, and therefore included into the text of the 1951 Peace Treaty a mandate to establish an international convention to ensure conservation of north Pacific fisheries resources. The International Convention for the High Seas Fisheries of the North Pacific Ocean was signed between Japan, Canada, and the United States in 1952, and was based on the abstention principle that grew out of U.S. fisheries policy of the 1930s and 1940s (VanCleve and

Johnson 1963, Atkinson 1964, Taguchi 1966, Johnson 1967, Kasahara and Burke 1973, Hollick 1981, Jackson and Royce 1986). The treaty established the International North Pacific Fisheries Commission (INPFC) to realize the objectives of the treaty and to serve as a forum for scientific deliberations, exchange of statistics, and publication of research relating to treaty issues.

Until 1978 the North Pacific Treaty regulated the Japanese high seas salmon fisheries only by setting their eastern boundary. Virtually nothing was known about ocean distribution of North American and Asian salmon at the time of treaty negotiation, so a provisional eastern boundary was set in an Annex under the treaty at 175°W in the Bering Sea and about 175°20'W south of Atka Island (Figure 2). Early INPFC-related research on high seas salmon entailed a wide variety of distributional and stock-determination studies, and by the late 1950s these studies showed that certain North American stocks (principally Bristol Bay sockeye salmon) migrate well west of 175°W and that certain Asian stocks (particularly chum salmon) migrate far east of the line (see early INPFC Annual Reports). Despite accumulating evidence that North American salmon were intercepted in large numbers by the mother-ship fishery, no consensus was ever reached in many years of INPFC deliberations regarding the qualifications of North American salmon stocks for abstention or the appropriateness of the 175°W boundary, so as a compromise the original treaty continued until 1978.

In 1977 the United States implemented the Fishery Conservation and Management Act of 1976 (FCMA), which established a 200-mile fishery conservation zone and claimed management authority over anadromous salmonids of U.S. origin throughout their migratory range (except where their range is within another country's 200-mile zone as recognized by the United States). The FCMA also called for withdrawal from international agreements that could not be renegotiated to be brought into conformity with the new law. Canada implemented similar legislation shortly thereafter. Renegotiation of the North Pacific Treaty took place in early 1978, and resulted in a revised Protocol and a new Annex that contained certain regulations for the Japanese high seas salmon fisheries. The new Annex closed the area southeast of 56°N, 175°E, prohibited the mother-ship fishery from operating before June (its season had begun in May in previous

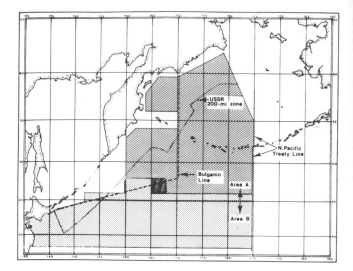

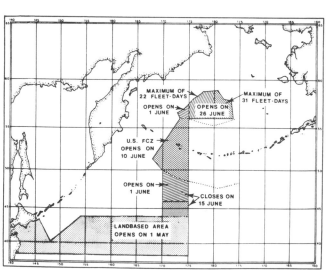

Figure 2. *Area of the Japanese mother-ship (cross-hatched) and land-based driftnet (stippled) salmon fisheries before 1977 (top panel) and in 1978-85 (bottom panel).*
Notes: The fishing area in 1977 was the same as shown in the top panel, except the U.S.S.R. 200-mile zone was excluded. The 15 June closure of the area 170°E-175°E and between 46°N and the U.S. 200-mile zone first went into effect in 1979.

years), set fishing periods for various areas inside and outside the U.S. 200-mile zone, and limited the number of mother-ship fleet-days northeast of 56°N, 175°E (see Figure 2). Most of these new regulations were designed specifically to protect Bristol Bay sockeye salmon.

In 1980 the mother-ship fishery, operating under regulations specified by the new North Pacific Treaty Annex, made the largest catch of chinook salmon in its history (704,000 fish, compared to an average of 250,000 fish in 1964-79 and a previous maximum of 554,000 fish in 1969). Most of the large chinook catch was made in the central Bering Sea, where previous research (summarized by Major et al. 1978) showed chinook salmon to be predominantly of western Alaskan and Canadian Yukon origin. This large catch fueled the concern of western Alaskan fishers that the 1978 revised treaty Annex gave inadequate protection to North American salmonids in general and to western Alaskan chinook salmon in particular.

In 1985 and 1986 the United States and Japan held a series of bilateral negotiations on the continuing salmon interception problem, and in March 1986 reached agreement on measures designed to reduce interceptions of North American salmon. One month later, all three INPFC member-nations met and agreed upon a revision of the Annex to the treaty. The new Annex incorporates the basic terms of the U.S.-Japan bilateral agreement and features a gradual phasing out of mother-ship fishing in the central Bering Sea (north of 56°N) through 1994, after which the area north of 56°N will be closed. Also, the new Annex repositioned the eastern boundary of the land-based fishery area at 174°E, although this boundary will be open to renegotiation before the 1991 season on the basis of results of scientific research designed to clarify further the continental origins of salmonids in the land-based fishery area.

U.S.S.R.-Japan Treaty

The U.S.S.R.-Japan Treaty was negotiated in 1956, as a result of U.S.S.R. concern about the new, rapidly growing Japanese high seas fisheries (Moiseev 1956, Atkinson 1964, Taguchi 1966). A major function of the Soviet-Japan Fisheries Commission (SJFC), created by the treaty, has been the setting of annual high seas salmon catch quotas, but it has also regulated times and areas of operation, gear characteristics, and other aspects of the fisheries. The total high seas salmon catch quotas have been steadily reduced from 120,000 mt in 1957 to 80,000 mt in 1976. A major blow to the fisheries came in 1977 when the U.S.S.R. further reduced the quota (to 62,000 mt) and established a 200-mile zone which was entirely closed to salmon fishing. In 1978 a further revised U.S.S.R.-Japan agreement closed the area northwest of 44°N, 170°E (except for a small section inside the U.S. 200-mile

zone), established a much more restrictive quota system, and obligated Japan to pay a cooperation fee for the privilege of catching salmon of U.S.S.R. origin on the high seas. The 1978-84 quotas were specified in terms of weight and numbers of fish, and quotas were set for the total ocean harvest as well as for the high seas catches made outside of Japanese and U.S. 200-mile zones. The total salmon quotas were further reduced, from 42,000 mt in 1978 to 40,000 mt in 1984. For the high seas area, there have also been separate species quotas. The quotas agreed upon by the U.S.S.R. and Japan have been allocated by the government of Japan to the various fishery components.

In 1985 a new 3-year agreement was signed, ostensibly based more on the principles embodied in the draft United Nations Law of the Sea than past U.S.S.R.-Japan agreements had been. The specific protocol regarding the Japanese high seas salmon fisheries featured a new quota system more restrictive than that of 1978-84. The 1985-86 quotas were set for each salmon species, in numbers and weight, for each of six major regions: (1) the mother-ship area north of the U.S. 200-mile zone; (2) the mother-ship area within the U.S. 200-mile zone; (3) the area between the U.S. 200-mile zone and 44°N; (4) the land-based fishery area south of 44°N and east of the U.S.S.R. and Japanese 200-mile zones; (5) the Japanese 200-mile zone in the north Pacific; and (6) the Japanese 200-mile zone in the Japan Sea. The total quota in 1985 was 37,600 mt, a slight reduction from 1984, but the quota for 1986 dropped significantly to 24,500 mt. The cooperation fees continued in 1985-86. The land-based fishery had previously begun operations in May, but in 1985 it did not begin until early June because of delay in agreement on the protocol. The 1986 protocol, however, explicitly gave the land-based fishery a 1 June to 31 July fishing season (although the fishery must terminate fishing when its quota is reached).

EFFECTS OF TREATY CHANGES ON FISHING EFFORT AND CATCHES

The level and pattern of fishing effort and catches by the Japanese high seas salmon fisheries changed greatly after the new treaty-related regulations were imposed in 1977-78. To illustrate these changes, catch and effort statistics reported through INPFC are analyzed with respect to two periods, 1972-76 and 1978-85. The year 1977 is not included in either period as it was a transitional year and the fishery area and timing were different from those in either period. Table 1 gives annual and mean 1972-85 fleet size, quota, fishing effort, and catch statistics for the mother-ship fishery, and Table 2 provides a similar compilation for the land-based fishery. For the examination of spatial distribution of effort and catch, the fishery areas are divided into 5°-longitude subareas as depicted in Figure 3. Subareas 1-10 for the mother-ship fishery area were defined by Fredin and Worlund (1974), and subareas 11-15 for the land-based area east of 160°E were defined by Myers et al. (1984). In addition, the land-based area west of 160°E is pooled into a single subarea for the purpose of this summary.

The mean spatial allocation of mother-ship and land-based driftnet fishing effort in 1972-76 and 1978-85 is shown in Figure 4, and the mean spatial distributions of catches of sockeye, chum, pink, coho, and chinook salmon are similarly shown in Figures 5-9, respectively. The mother-ship fishery has not retained incidentally caught steelhead (except for scientific sampling), and therefore no

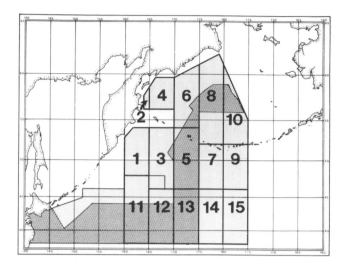

Figure 3. The high seas salmon fishery area and statistical subareas used in this study.

Note: For the purpose of compiling land-based driftnet statistics, subarea 12 is considered to extend north to 48°N.

Table 1. Japanese mother-ship salmon fishery effort and catch, 1972-85.

Year	No. of fleets	No. of catcher boats	Fishing effort (1000s of tans[a])	Catch quota (mt)	Total catch (mt)	Catch (In 1000s of fish) Total	Sockeye	Chum	Pink	Coho	Chinook
1972	10	332	5,917	35,236	35,205	21,227	3,184	13,373	3,795	615	261
1973	10	332	5,850	35,732	35,588	23,597	2,613	7,857	12,018	989	119
1974	10	332	5,433	33,702	33,563	20,767	2,282	9,283	7,756	1,085	361
1975	10	332	5,633	34,108	33,908	24,709	2,171	7,367	14,654	356	162
1976	10	332	5,811	32,484	32,418	21,048	2,277	10,444	7,215	827	285
1972-76 mean:			5,729	34,252	34,136	22,270	2,505	9,665	9,088	774	238
1977	6	245	3,984	23,957	23,565	16,778	1,508	5,996	9,100	79	93
1978	4	172	2,721	15,500	15,399	8,251	1,882	3,802	1,853	609	105
1979	4	172	2,798	15,500	15,450	9,275	2,186	3,277	3,405	281	126
1980	4	172	3,146	15,500	15,442	7,431	2,412	3,098	561	656	704
1981	4	172	2,902	15,500	15,445	9,560	2,224	2,539	4,094	615	88
1982	4	172	2,942	15,500	15,429	7,899	1,738	3,217	1,654	1,183	107
1983	4	172	2,954	15,500	15,414	9,445	1,655	3,081	4,324	297	87
1984	4	172	2,740	14,600	14,522	7,170	1,596	3,276	1,430	786	82
1985	4	172	2,322	14,080	12,466	6,885	1,139	2,836	2,717	128	66
1978-85 mean:			2,816	15,210	14,946	8,240	1,854	3,141	2,505	569	171[b]

Sources: Data on fleet size are from Fredin et al. (1977) and INPFC Documents. Information on quota allocations to the mother-ship fishery is from Fredin et al. (1977) and M.L. Dahlberg (U.S. National Marine Fisheries Service, personal communication). Effort and catch statistics are from INPFC (1979), INPFC Statistical Yearbooks, or INPFC Documents.

[a] A tan is approximately 47 m (154 ft) long.

[b] Mean excluding the anomalous 1980 values is 94.

Table 2. Japanese land-based driftnet salmon fishery effort and catch, 1972-85.

Year	No. of vessels	Fishing effort		Catch quota[a] (mt)	Total catch[a] (mt)	Catch (In 1000s of fish)						
		Vessel-days	1000s of tans			Total	Sockeye	Chum	Pink	Coho	Chinook	Steelhead
1972	374	13,517	5,154	46,774	(45,802)	29,672	3,711	8,598	14,839	2,421	103	—
1973	374	15,087	5,754	50,068	49,854	35,527	3,308	7,614	20,650	3,794	162	—
1974	374	15,784	6,020	44,498	44,057	30,321	3,155	12,179	11,242	3,559	186	—
1975	371	15,709	5,991	47,892	47,517	33,481	3,969	11,479	15,347	3,550	135	—
1976	368	15,590	5,946	42,916	40,645	27,120	3,504	11,432	9,235	2,751	198	—
1972-76 mean:		15,137	5,773	46,430	45,575	31,224	3,529	10,260	14,263	3,215	157	—
1977	298	9,751	3,719	34,443	32,569	24,428	1,289	6,230	15,041	1,722	146	—
1978	209	8,841	3,373	20,600	20,425	15,349	1,293	3,488	7,846	2,512	210	—
1979	209	7,545	3,218	20,600	20,450	15,968	757	2,661	11,190	1,199	162	—
1980	209	8,500	3,144	20,600	20,016	16,461	787	2,697	11,612	1,205	160	—
1981	209	9,243	3,234	20,600	19,956	16,082	859	2,509	11,292	1,209	190	(23)
1982	209	8,192	2,962	20,600	20,229	16,081	723	2,930	11,035	1,201	165	27
1983	209	8,715	3,114	20,600	20,192	15,860	828	2,395	11,308	1,122	178	29
1984	209	7,858	2,824	19,400	14,421	13,248	305	2,214	9,727	894	92	15
1985	209	6,842	2,442	17,115	15,141	12,441	155	1,432	9,973	766	101	14
1978-85 mean:		8,217	3,039	20,014	18,854	15,186	713	2,541	10,498	1,264	157	20

Sources: Information on fleet size and effort in vessel-days is from Japan Fisheries Agency. Information on quotas for 1972-77 is from Fredin et al. (1977) (it is uncertain whether these quotas include the small-vessel component), and information on large-vessel quotas for 1978-85 is from M.L. Dahlberg (U.S. National Marine Fisheries Service, personal communication). Catches in weight are from INPFC (1979) or INPFC Statistical Yearbooks (drafts for 1984-85); the 1972 value is estimated by applying mean weights in Table 92 of INPFC (1979) to catches reported in numbers of fish. Catches in numbers are from INPFC Statistical Yearbooks or Documents. Steelhead catch statistics are not available before 1981; the 1981 value is an estimate for the entire fleet based on 139 reporting vessels.

[a] Does not include steelhead.

mother-ship catch statistics for steelhead are available. The land-based driftnet fishery has retained steelhead in recent years, and Japan began to provide land-based catch statistics for steelhead in 1981. Figures 4-9 show mean absolute amount of effort or catch, coded according to unequal intervals considered most appropriate given the spread of values. The values of mean effort for the land-based fishery are not exactly comparable with those shown for the mother-ship fishery because tans of nets used by the land-based fishery vary somewhat in length. The mean monthly distributions of fishing effort and catches in 1972-76 and 1978-85 in the two fisheries are presented in Table 3. Statistics representing a small amount of out-of-area fishing by the early mother-ship fishery and recent land-based fishery were excluded from the compilation.

Mother-ship fishery

Catch quotas for the mother-ship fishery were reduced gradually throughout the 1960s to mid-1970s, but in 1977 and 1978 there were substantial quota reductions (see Table 1). The mean quota in the recent period is less than half of the mean for 1972-76. The reduction of catch quotas coupled with the major reduction of fishery area in 1977-78 necessitated roughly commensurate reductions of fleet size and fishing effort. The close correspondence between quotas and total reported catches is striking.

With few exceptions the catches of all species have been less in 1978-85 than in 1972-76. The total catch in numbers has declined more than the total catch in weight, due mainly to the especially large reduction in catch of pink salmon, which are considerably smaller than the other four species. The declines in pink and chum salmon catches have been greatest, while catches of sockeye and coho salmon have declined only modestly. Trends in the chinook salmon catch are more difficult to discern due to the anomalous abundance of the species in 1980; 1978-85 catches have averaged about 40 percent of the 1972-76 catches if the unusually large 1980 catch is not considered. The smaller reduction of sockeye catch relative to pink and chum salmon catches is undoubtedly an important factor in the recent economics of the

Table 3. Mean temporal allocation of fishing effort and catch by species by the Japanese mother-ship and land-based driftnet salmon fisheries, 1972-76 and 1978-85.

			Effort	Sockeye	Chum	Pink	Coho	Chinook
Mother-ship fishery								
1972-76		May	17	31	11	6	[a]	3
		June	47	39	29	55	6	24
		July	36	30	60	39	94	73
1978-85		May	0	0	0	0	0	0
		June	51	46	46	51	1	14
		July	49	54	54	49	99	86
Land-based fishery								
1972-76		May	47	82	54	21	4	23
		June	34	13	40	73	25	31
		July	19	5	6	6	71	46
1978-85		May	36	65	38	16	5	25
		June	48	31	50	63	63	53
		July	16	3	12	21	32	22

Note: Values are mean percent of total annual effort (in tans of gillnet) or catch (in numbers of fish) that occurred in each month.
[a] Less than 0.5%.

fishery. The proportion of sockeye salmon in the total mother-ship catch in numbers doubled (from an average of 11 percent in 1972-76 to 23 percent in 1978-85). Because sockeye salmon are generally more valuable in the market than chum or pink salmon (Herrfurth 1985, Ignel and Dahlberg 1985), this change signifies a higher relative product value in recent years.

Figure 4 and Table 3 depict how fishing effort was distributed in time and space in 1972-76 and 1978-85. An average of 17 percent of the total annual effort was expended in May, and June was more heavily fished than July, in 1972-76, whereas during the more recent period there was no fishing in May and effort levels in June and July were about equal. In 1972-76, effort was distributed throughout the fishery area, heaviest fishing having been in subareas 3, 5, 7, 8, and 10. In 1978-85, fishing effort was highly concentrated in subarea 5. Bering Sea subareas 8 and 10 have not opened until late June in the recent period, which partially explains the recent low effort there. However, Dahlberg (National Marine Fisheries Service, personal communication) showed that the number of fleet-days expended in subareas 8 and 10 was always less in the recent period than permitted by the fleet-day limits imposed by the 1978 North Pacific Treaty Annex.

The changes in pattern of fishing had a major effect on the temporal and spatial pattern of sockeye salmon catch. Maturing sockeye (those destined to spawn in the year of the ocean fishery) migrate through the fishery area in May and June, but by late June the great majority of them are in nearshore waters off Asia and North America (French et al. 1976). The migratory patterns of immature sockeye, on the other hand, make them most available in the fishery area in June and especially July. In 1972-76 sockeye were caught in all subareas but highest catches occurred in subareas 3 and 5 and to a lesser extent in subarea 7 (see Figure 5). Nearly equal proportions of the total annual catch were obtained in May, June, and July. In the recent period, the great majority of the catch has been made in subarea 5, and slightly over one-half of the catch has been made in July.

In 1972-76 the largest catches of chum salmon were made in the central Bering Sea (mostly in the northern parts of subareas 8 and 10), and smaller but still significant catches were made in subareas 3 and 5 (see Figure 6). Catches of chum increased throughout the season, and, on the average, over one-half of the annual catch was made in July. In the recent period, the great majority of the chum catch was made in subarea 5, and slightly over one-half of the catch was made in July, as is the case for sockeye.

Pink salmon tend to be most abundant in the western sectors of the fishery area, largely because pink salmon in the mother-ship area are nearly all of Asian origin (Takagi et al. 1981), and therefore the homeward migrations tend to shift principal

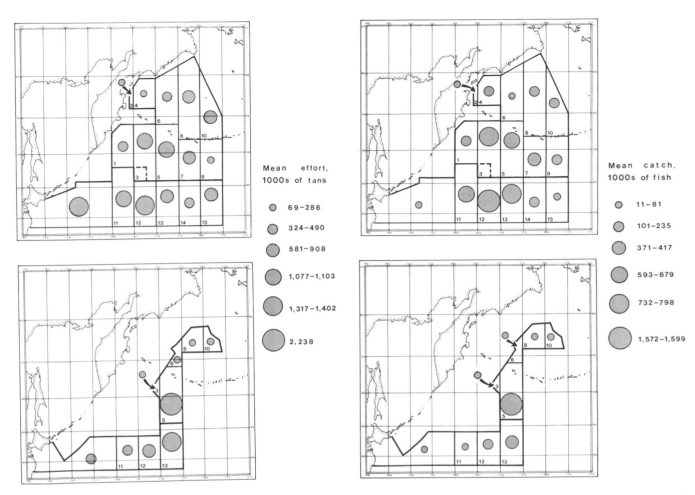

Figure 4. *Mean spatial distribution of fishing effort by the Japanese mother-ship and land-based driftnet salmon fisheries in 1972-76 (upper panel) and 1978-85 (lower panel).*
Notes: For each statistical subarea, the average annual effort (in thousands of tans of drift gillnet) was calculated and represented by shaded circles according to ranges shown. The land-based effort values are not exactly comparable to mother-ship effort values (see text).

Figure 5. *Mean spatial distribution of sockeye salmon catch by the Japanese mother-ship and land-based driftnet salmon fisheries in 1972-76 (upper panel) and 1978-85 (lower panel).*
Note: For each statistical subarea, the average annual catch (in thousands of fish) was calculated, and represented by shaded circles according to ranges shown.

concentrations westward during the late spring and summer months. Consequently, the mother-ship fishery in 1972-76 made its greatest pink salmon catches in the western and especially southwestern subareas 1, 3, 4, 5 and 7 (see Figure 7). Only a small fraction of the total catch was made in May, and slightly over one-half of the annual catch was, on the average, made in June. Recently, most of the catch has come from subarea 5, about equally from June and July.

Coho salmon are most abundant in the southeastern sectors (mainly subareas 5, 7 and 9) in July; abundance in the other subareas or before July is usually very low. In 1972-76 coho catches were greatest in subareas 5, 7 and 9 in July, while in the recent period virtually all of the coho catch has come from subarea 5 in July (see Figure 8, Table 3).

The pattern of chinook salmon catch in 1972-76 was generally similar to that of chum salmon. Largest catches were made in subareas 8 and 10 in July. Recently, the fraction of catch made in subarea 5 has been much greater than before, due to the heavy level of fishing there. A higher fraction of the catch has been made in July in recent years than in 1972-76.

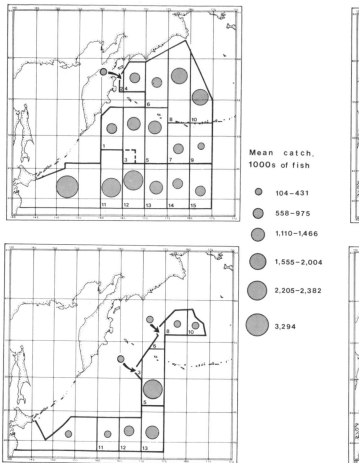

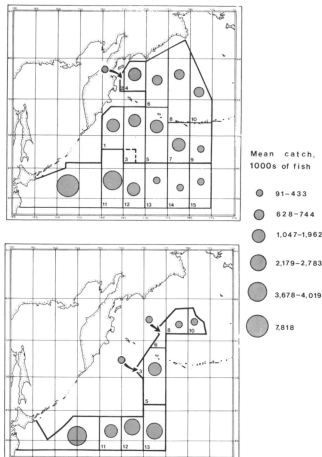

Figure 6. Mean spatial distribution of chum salmon catch by the Japanese mother-ship and land-based driftnet salmon fisheries in 1972-76 (upper panel) and 1978-85 (lower panel). See note to Figure 5.

Figure 7. Mean spatial distribution of pink salmon catch by the Japanese mother-ship and land-based driftnet salmon fisheries in 1972-76 (upper panel) and 1978-85 (lower panel). See note to Figure 5.

Land-based driftnet fishery

Quotas allocated to the land-based driftnet fishery in recent years have been less than one-half of those in 1972-76, and total reported catches have declined commensurately (see Table 2). Declines in catch were greatest for sockeye and chum salmon, the recent catches of these species averaging only about 20 percent and 25 percent, respectively, of the 1972-76 catches. The pink salmon catch, by contrast, averaged 74 percent of the 1972-76 period, and the species has recently constituted about 69 percent of the total catch in numbers, compared to about 46 percent in 1972-76. The much higher proportion of pink salmon in recent catches accounts for the decline in total land-based catch in weight having been greater than the decline in

numbers of fish (see Table 2). The catch of coho salmon has declined by about 61 percent, but the mean chinook salmon catch has not changed.

In 1972-76 fishing effort by the land-based fishery was heaviest west of 170°E (see Figure 4). Nearly one-half of the total effort occurred in May, about one-third was expended in June, and July was lightly fished (see Table 3). Harris (1978) examined the 1972-76 pattern of fishing, and found that effort in May was concentrated in the middle sector (subareas 11-13) of the fishery area, mainly north of 44°N, where sockeye salmon were in greatest abundance in the fishery area. In June, effort shifted heavily westward, where pink salmon especially but also chum salmon were most abundant. By July, coho became abundant in the eastern sector

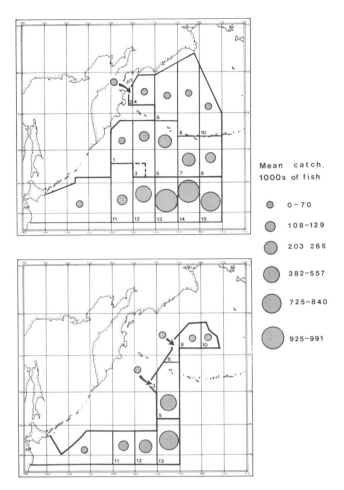

Mean catch,
1000s of fish

- 0 – 70
- 108 – 129
- 203 265
- 382 557
- 725 – 840
- 925 – 991

Figure 8. Mean spatial distribution of coho salmon catch by the Japanese mother-ship and land-based driftnet salmon fisheries in 1972-76 (upper panel) and 1978-85 (lower panel). See note to Figure 5.

of the fishery area (subareas 13-15) and effort (albeit much less than in May and June) shifted far to the east. The recent pattern of fishing has been very different. Effort in 1978-85 has been highest in subareas 12 and especially 13, and there has been a reversal in the relative amounts of fishing effort expended in May and June. The area west of 160°E has recently been fished far less heavily than before.

The land-based sockeye catch has declined greatly, largely because most of the area where sockeye had been caught in 1972-76 was excised from the fishery (Harris 1987). About 99 percent of the 1972-76 sockeye catch had been made north of 44°N, while in 1978-85 fishing north of 44°N has been permitted only in subarea 13 through mid-June (see Figure 2). Recently, then, the great-

est part of the sockeye catch has been obtained in subarea 13. Sockeye have still been caught mostly in May, but a higher fraction of the catch has been made in June in recent years.

The spatial and temporal pattern of chum salmon catch by the land-based fishery has also changed markedly (see Figure 6, Table 3). Most of the catch was made west of 170°E in 1972-76, while recently the major fraction has been obtained in subarea 13. Most of the catch has been made in May and June, as before, but the fraction obtained in June has increased.

In 1972-76 the land-based fishery made its greatest catches of pink salmon west of 165°E in June. Recently the catch has been rather evenly distributed across the fishery area (see Figure 7), but still predominantly from June fishing.

As mentioned above, coho salmon are distributed mostly in the eastern sectors of the land-based fishery area, and previously important coho fishing grounds were eliminated by the closure of the region east of 175°E in 1978 (east of 174°E beginning in 1986). Subarea 13 has recently provided most of the coho catch (see Figure 8). The temporal pattern of coho catch has shifted somewhat earlier; a much higher proportion of the annual catch has recently been made in June (see Table 3).

The only notable change in the pattern of chinook catch by the land-based fishery is that a higher proportion of the annual catch has recently been made in June than in 1972-76. There is no striking shift in spatial distribution of the catch within the area remaining open to the fishery (see Figure 9).

EFFECTS OF REGULATORY CHANGES ON HIGH SEAS CATCHES OF NORTH AMERICAN SALMONIDS

The numbers of salmonids of North American origin caught by the Japanese high seas salmon fisheries are influenced not only by the overall level of catch, but also by the spatial and temporal distribution of catch relative to the migratory patterns of North American fish in the fishery area. Estimating high seas interceptions of North American salmonids entails deriving estimates of the proportion of North American fish in various temporal and spatial strata and applying these to similarly stratified catch statistics.

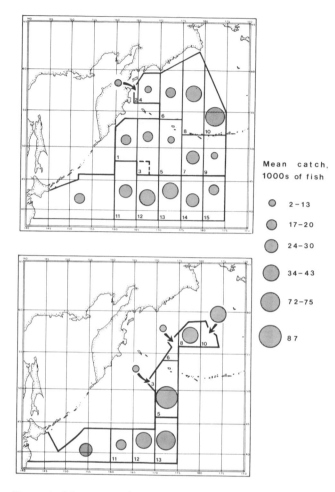

Mean catch,
1000s of fish

- 2–13
- 17–20
- 24–30
- 34–43
- 72–75
- 87

Figure 9. Mean spatial distribution of chinook salmon catch by the Japanese mother-ship and land-based driftnet salmon fisheries in 1972-76 (upper panel) and 1978-85 (lower panel). See note to Figure 5.

pattern, genetic, and morphometric studies) are more conducive to quantitative use, but they inherently have some degree of uncertainty.

Information gained from INPFC-related studies through the early to mid-1970s is summarized in INPFC comprehensive reports as follows: French et al. (1976) for sockeye, Neave et al. (1976) for chum, Takagi et al. (1981) for pink, Godfrey et al. (1975) for coho, and Major et al. (1978) for chinook salmon. A similar INPFC comprehensive report on origins and distribution of steelhead (including North American *Salmo gairdneri* as well as a closely related Asian anadromous trout, *S. mykiss*) is in preparation. Much of the research done since these reports were prepared was in response to the research mandate of the 1978 revised North Pacific Treaty, and pertained mainly to the land-based fishery area and southern mother-ship fishery area.

The Fisheries Research Institute, University of Washington, is engaged in a study to estimate catches of North American salmonids by the Japanese high seas salmon fisheries, and to assess the effectiveness of regulatory provisions in the 1978 and 1986 North Pacific Treaty Annexes in reducing interceptions of North American fish. To date, quantitative estimation of 1972-84 high seas interceptions of North American sockeye salmon has been completed and reported by Harris (1987); analysis for the other salmon species and steelhead is still in progress. For the purpose of this paper, I summarize the results and conclusions of Harris (1987) for sockeye, and summarize information for other species in a more qualitative manner.

Virtually all available information on continental origins of high seas salmonids has resulted from INPFC-related research programs of Canada, Japan, and the United States. Beginning in 1955, there have been numerous diverse studies including extensive tagging studies, general distributional studies, and studies employing serological, parasitological, or morphometric analysis, scale pattern recognition, analysis of age and maturity composition, and other techniques to estimate stock origins. Generally, tagging and parasitological studies can provide proof of occurrence of certain stocks or stock groups in a high seas area, but such data are often slow to accumulate and are not conducive to quantitative estimation of stock composition. Data from statistical studies (such as scale

Sockeye salmon

The known ranges of Asian and North American sockeye salmon (immature and maturing fish combined) in May-July between 160°E and 160°W are shown in Figure 10. For each 2°-latitude by 5°-longitude area, the figure shows the numbers of Asian and/or North American coastal tag recoveries from 1956-85 releases in the area. Virtually all of the Asian recoveries have been made on the Kamchatka Peninsula, which produces the great majority of Asian sockeye. Most of the North American recoveries have been made in western Alaska (particularly Bristol Bay), but from releases east of 175°E, west of 175°W, and south of the Aleutian Islands there have been seventeen recoveries from central Alaska and eight from southeast Alaska/ British Columbia. Occurrence of Kamchatkan or Bristol Bay fish in certain areas for which there are no tagging data is known from a study of region-specific parasites (Margolis 1963); these additional known occurrences are indicated in Figure 10 by the letter "P." Western Alaskan sockeye clearly range most of the way across the pre-1977 mother-ship fishery area and into at least the northeast corner of the land-based area. As mentioned above, central Alaskan and southeast Alaskan/British Columbian sockeye range at least into subareas 7 and 9. Asian sockeye range across the entire high seas fishery area, although known occurrences in the eastern sectors are spotty.

Estimates of interception of western and central Alaskan sockeye by the mother-ship and land-based fisheries are presented in Tables 4 and 5, respectively.[1] The numbers of southeast Alaskan and British Columbian sockeye caught east of 175°E in 1972-77 were likely very small, and were not estimated. The scope of the original scale pattern study of Cook et al. (1981) precluded estimation of stock composition for sockeye in minor age groups or for sockeye caught in July or west of 160°E. Therefore, the estimates presented in Table 5 pertain to "effective" catches within the scope of the scale study, which averaged 80 percent of the total reported land-based catches in 1972-84.

The mother-ship fishery's catch of maturing sockeye in 1978-84 averaged 885,000 fish, less than

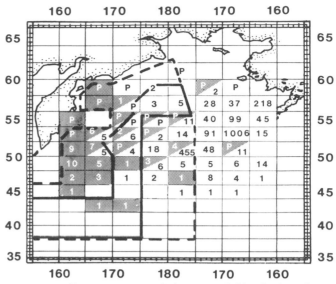

Figure 10. *Known ranges of Asian[1] and North American[2] sockeye salmon between 160°E and 160°W, in May-July, immature and maturing fish combined.*
[1] Partially or fully shaded areas.
[2] Partially or fully open areas containing data.
Notes: Data are the numbers of Asian and/or North American coastal recoveries from tag releases in the areas (1956-85 INPFC-related tagging). "P" signifies known presence of Kamchatkan or Bristol Bay sockeye on the basis of parasitological evidence (Margolis 1963) in areas where the stock-group is not known to occur by tagging data. The dotted and solid bold lines outline the pre-1977 and 1978-84 high seas salmon fishery areas, respectively.

one-half of the 1972-76 level, while the recent catch of immature fish has greatly increased from an average of 642,000 fish in 1972-76 to an average of 1,071,000 fish (see Table 4). This change toward a higher percentage of immature fish is due to the temporal shift of sockeye catch caused in turn by recent prohibition of May fishing and heavy concentration of fishing effort in subarea 5 in July. The proportion of sockeye of Alaskan origin in the catch of maturing fish declined from an average of 36 percent in 1972-76 to 28 percent in 1978-84. These Alaskan fish were virtually all of western Alaskan origin, although small numbers of central Alaskan fish were caught in subareas 7 and 9 in 1972-77 (Harris 1987). The catch of maturing fish of Alaskan origin decreased from an average of 674,000 fish in 1972-76 to 244,000 fish in 1978-84. The proportion of fish of Alaskan origin in the catch of immature sockeye decreased also, from an average of 50 percent in 1972-76 to 36 percent in 1978-84; because of the greatly increased catches of immatures, however, the average catch of immatures of Alaskan origin actually increased, from

---

[1] These estimates of catches of North American sockeye by the mother-ship and land-based fisheries are from Harris (1987). They were calculated by a modification of the methods of Fredin and Worlund (1974), who estimated catches of Bristol Bay sockeye by the mother-ship fishery in 1956-70, and of Meyer and Harris (1983), who estimated catches of North American sockeye by the land-based fishery in 1972-81. Fredin and Worlund (1974) initially analyzed tagging, parasitological, and other data to identify areas and times of major intermingling of Asian and Bristol Bay sockeye, and to devise a method for estimating maturity composition of the mother-ship catches. They then used data on age composition of the mother-ship catches to estimate numbers of Bristol Bay sockeye caught in times and areas of assumed intermingling. Harris (1987) used this general method, but considered that the intermingling of Asian and Bristol Bay sockeye was wider, in time and space, than assumed by Fredin and Worlund. Also, estimates were made of mother-ship catches of western and central, and not only Bristol Bay, Alaskan sockeye. Meyer and Harris (1983) used the results of scale pattern analyses by Cook et al. (1981) and Knudsen and Harris (1982) in conjunction with age and maturity composition data and land-based fishery catch statistics to estimate the fishery's interceptions of North American fish in 1972-81. Harris (1987) used Meyer and Harris's estimates for 1972-81, and applied their averaged stock composition estimates to recent age and maturity data to update estimates for 1982-84.

Table 4. Estimated catches of western and central Alaskan sockeye salmon by the Japanese mother-ship salmon fishery, 1972-84.

	Maturing fish			Immature fish			Total		
	Total catch (In 1000s of fish)	Catch of Alaskan origin	Percent of Alaskan origin (%)	Total catch (In 1000s of fish)	Catch of Alaskan origin	Percent of Alaskan origin (%)	Catch (In 1000s of fish)	Catch of Alaskan origin	Percent of Alaskan origin (%)
1972	2,346	899	38	835	254	30	3,181	1,153	36
1973	2,133	783	37	475	260	55	2,608	1,043	40
1974	1,333	357	27	947	672	71	2,280	1,029	45
1975	1,696	555	33	471	216	46	2,167	771	36
1976	1,794	778	43	482	233	48	2,276	1,011	44
1977	1,167	546	47	341	320	94	1,508	867	58
1978	862	232	27	1,020	298	29	1,882	530	28
1979	907	188	21	1,278	413	32	2,185	601	28
1980	1,130	298	26	1,282	587	46	2,412	885	37
1981	744	223	30	1,479	445	30	2,223	668	30
1982	773	220	28	965	303	31	1,738	523	30
1983	784	244	31	871	269	31	1,655	513	31
1984	998	301	30	598	330	55	1,596	631	40
Mean, 1972-76	1,860	674	36	642	327	50	2,502	1,001	40
Mean, 1978-84	885	244	28	1,071	378	36	1,956	622	32

Source: Harris (1987).

327,000 fish to 378,000 fish. Virtually all of these Alaskan immatures were of western Alaskan origin.

The decline in the proportion of fish of Alaskan origin in catches of maturing and immature sockeye was not as great as might be expected given the closure southeast of 56°N, 175°E. However, the U.S.S.R.-Japan closure of the sector northwest of 44°N, 170°E (except for a small part of subarea 3) in 1978 tended to push the fishery eastward, offsetting effects of the North Pacific Treaty Annex. This can be illustrated by examining trends in the mean longitude (calculated from statistics coded by 5°-longitude interval) of sockeye catch. In 1972-76 it was 169.2°E (range 168.2°E to 170.3°E), while in 1978-84 it was 170.4°E (range 169.9°E to 171.0°E). Thus, despite the closure southeast of 56°N, 175°E, the recent mother-ship sockeye catches have been made, on the average, slightly closer to North America than before. The mean longitude of sockeye catch in 1977 (when the U.S.S.R. 200-mile zone was closed but the sector between 175°E and 175°W was still open) was 171.8°E, farther east than in any other year in the 1972-84 period, and the estimated composition of fish of Alaskan origin

in the 1977 mother-ship catch was higher than in any other year (see Table 4).

The land-based fishery's catches of sockeye were greatly reduced by the regulatory regime imposed in 1977-78. In marked contrast with trends in the mother-ship catches, the land-based fishery's catches of both maturing and immature fish greatly declined (see Table 5). The lack of a sharp shift in maturity composition of the catches is likely attributable to the fact that the recent catches have still been made mainly in the early part of the fishing season. The proportions of Alaskan origin in the catches of both maturing and immature sockeye essentially did not change after 1977-78. The U.S.S.R.-Japan agreements had the effect of reducing overall sockeye catch and limiting the catch to south of 44°N (except in subarea 13 through mid-June), but they had little effect on the longitudinal or temporal distribution of the catches; most of the recent sockeye catches were made early in the season in subareas 11, 12 and 13, as before. Although there was a latitudinal shift in sockeye catch, it probably had little effect on continental stock composition of the catch (Cook et al. 1981, Meyer and

Harris 1983). The North Pacific Treaty Annex, which closed the sector east of 175°E, also had little effect on continental stock composition of the land-based sockeye catches. Although the proportion of fish of Alaskan origin between 175°E and 175°W is higher than west of 175°E (Meyer and Harris 1983), an average of only 9 percent of the annual sockeye catch was made east of 175°E in 1972-77. Therefore, exclusion of the sector had little effect on stock composition of the total catches.

In summary, the revised 1978 North Pacific Treaty Annex had the desired effect of protecting Alaskan sockeye from high seas catch more effectively than the original Annex, but the actual savings of Alaskan fish in recent years are due to the combined effects of the revised U.S.S.R.-Japan Treaty and North Pacific Treaty. Certainly the overall reduction in total mother-ship sockeye catches is principally due to lowered catch quotas and the concomitant decrease in fleet size and fishing effort. The decline in numbers of Alaskan maturing fish caught by the mother-ship fishery can be attributed largely to prohibition of May fishing

and closure of the sector southeast of 56°N, 175°E. While the recent mother-ship catches of immature sockeye increased greatly over 1972-77 levels, the recent catch of immatures of Alaskan origin has only slightly increased due to closure of the sector southeast of 56°N, 175°E. In the case of the land-based fishery, the continental stock composition of the catches essentially did not change; the greatly reduced catches of fish of Alaskan origin can be attributed to overall catch reductions caused in turn by significant removal from the fishery of the fishing area north of 44°N. The closure southeast of 56°N, 175°E served to prevent potential catches of Alaskan sockeye, as it precluded a major eastward shift of mother-ship and land-based fishing that otherwise might have occurred following the U.S.S.R.-Japan closure northwest of 44°N, 170°E.

The analysis by Harris (1987) is based on numerous assumptions, including those pertaining to the various studies summarized. One important assumption is that statistics reported for the mother-ship and land-based fisheries are accurate. Another noteworthy assumption is that regional stock

Table 5. *Estimated catches by the Japanese land-based driftnet fishery of western and central Alaskan sockeye salmon in major age groups, east of 160°E in May and June, 1972-84.*

	Maturing fish			Immature fish			Total			
	Effective total catch (In 1000s of fish)	Catch of Alaskan origin	Percent of Alaskan origin	Effective total catch (In 1000s of fish)	Catch of Alaskan origin	Percent of Alaskan origin	Total catch (In 1000s of fish)	Effective total catch	Catch of Alaskan origin	Percent of Alaskan origin
1972	988	95	10	1,624	196	12	3,711	2,612	291	11
1973	1,540	81	5	994	128	13	3,308	2,534	209	8
1974	1,207	77	6	1,166	152	13	3,155	2,376	229	10
1975	1,804	58	3	734	72	10	2,968	2,538	130	5
1976	1,772	318	18	909	76	8	3,504	2,681	393	15
1977	341	36	11	639	47	7	1,289	980	83	8
1978	308	29	9	705	56	8	1,293	1,013	85	8
1979	401	27	7	261	23	9	757	662	50	7
1980	225	21	9	382	37	10	787	607	59	10
1981	385	36	9	334	33	10	858	718	70	10
1982	526	55	10	130	15	12	723	657	70	11
1983	423	56	13	252	20	8	828	676	76	11
1984	129	13	10	116	8	7	305	245	21	9
Mean, 1972-76	1,462	126	8	1,085	125	11	3,329	2,548	250	10
Mean, 1978-84	342	34	10	312	28	9	793	654	62	9

Notes: In all cases, "effective total catch" includes only catches made east of 160°E in May and June; for maturing fish, only fish of ages 1.2, 2.2, 1.3, and 2.3, and for immatures, only fish of ages 1.2 and 2.2, are included in the effective total.
Source: Harris (1987).

composition of the sockeye population in the western north Pacific was generally the same in 1978-84 as in 1972-76. Because there is little recent information on stock origins, estimates for the later period used averaged values from studies based on earlier data. Given the great increase in abundance of Alaskan sockeye runs in recent years, however (Rogers 1984), these estimates of recent interceptions may be conservative.

Chum salmon

There is not as much information on origins of chum salmon in the pre-1977 high seas fishery area as for sockeye salmon. Available information comes mainly from 1956-85 high seas tagging experiments, and scale pattern analyses of maturing chum done recently by Ishida et al. (1983, 1984, 1985). The distribution of tag recoveries from May-July releases of tagged maturing and immature chum between 160°E and 160°W (Figure 11) suggests that the range of North American chum south of the Aleutian Islands overlaps only the eastern sector (175°E to 175°W) of the pre-1978 mother-ship and land-based fishery areas. Within this area of intermingling, there are far more tag returns from Asia, suggesting that Asian chum greatly predominate. Most of the North American recoveries from releases west of 175°W were from western Alaska. In the Bering Sea, North American chum occur at least in the eastern sector of the mother-ship fishery area (subarea 10), but there is one western Alaskan tag recovery from a release near 60°N, 175°E (see Figure 11). Although it is possible that North American chum range across much of the pre-1977 mother-ship fishery area in the Bering Sea, the numbers of Asian versus North American tag recoveries suggest that chum of Asian origin greatly predominate.

Recent scale pattern analyses also suggest that Asian chum predominate in most areas of the western Pacific Ocean and Bering Sea. Some fish sampled in the far western north Pacific and even Okhotsk Sea were classified to North American categories, however. The results of the scale pattern studies were considered by the authors to be questionable, due to factors relating to methodology, sample representativeness, and inadequate separability of defined stock groups, and work is currently under way to improve application of the

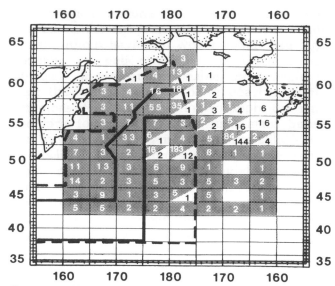

Figure 11. Known ranges of Asian[1] and North American[2] chum salmon between 160°E and 160°W, in May-July, immature and maturing fish combined.

[1] Partially or fully shaded areas.

[2] Partially or fully open areas containing data.

Notes: Data are the numbers of Asian and/or North American coastal recoveries from tag releases in the areas (1956-85 INPFC-related tagging). The dotted and solid bold lines outline the pre-1977 and 1978-85 high seas salmon fishery areas, respectively.

technique (Japan Fisheries Agency, personal communication).

On the basis of tagging data alone, I provisionally conclude that the mother-ship and land-based driftnet fisheries probably never caught substantial numbers of North American chum. This general conclusion was also reached by Fredin et al. (1977), who estimated, also on the basis of tagging data, that chum are entirely of Asian origin in all high seas salmon fishery areas except in the area between 50°-52°N, 175°E-175°W, where 4 percent are of Alaskan origin, and in subareas 8 and 10, where 5 percent are of Alaskan origin. The single North American recovery from releases in the land-based fishery area (see Figure 11) was not on record when Fredin et al. (1977) made their analysis but it would not change the conclusion that the great majority of the land-based catches of chum, even those east of 175°E, has been of Asian fish.

The closure of the sector southeast of 56°N, 175°E virtually eliminated the likelihood of any major interception of North American chum by the mother-ship and land-based driftnet fisheries in all areas but the central Bering Sea. In the Bering

Sea, the interceptions of North American chum would likely have been much less in recent years than before 1977 due to the overall decrease in chum catch in subareas 8 and 10 (see Figure 6).

Pink salmon

Information on regional origins of pink salmon in the western north Pacific and Bering Sea is mainly from coastal recovery of high seas tags (summarized in Figure 12) and from scale pattern and meristic studies, summarized by Takagi et al. (1981). The scale pattern and meristic studies, however, were not conclusive due to lack of sufficient separation between certain Asian and Alaskan stocks. The results of pink salmon tagging are generally similar to those for chum: North American fish are indicated to occur only in the eastern sectors of the pre-1977 mother-ship and land-based fishery areas, and Asian fish likely predominate in the areas of intermingling. All of the North American tag recoveries from releases west of 170°W have been from western Alaska. Owing to the (nearly) invariable 2-year life cycle of pink salmon, there are major differences in abundance between the even-

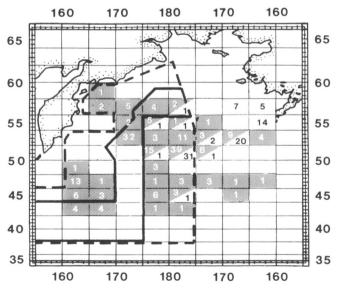

Figure 12. Known ranges of Asian[1] and North American[2] pink salmon between 160°E and 160°W, in May-July.

[1] Partially or fully shaded areas.

[2] Partially or fully open areas containing data.

Notes: Data are the numbers of Asian and/or North American coastal recoveries from tag releases in the areas (1956-85 INPFC-related tagging). The dotted and solid bold lines outline the pre-1977 and 1978-85 high seas salmon fishery areas, respectively.

numbered and odd-numbered year runs within certain regions. Asian stocks in general tend to be in greatest abundance in odd years, while the odd-year runs to western Alaska are usually negligible. Therefore, interception of North American pink salmon by the Japanese salmon fisheries would have occurred mainly in even-numbered years. Fredin et al. (1977) used tagging data to estimate interceptions of western Alaskan pink salmon in 1956-75, and the mean even-year estimate was 48,000 fish (range 0 to 253,000) and the mean odd-year estimate was only 2,000 fish (range 0 to 5,000).

As is the case for chum, the elimination of the fishing area southeast of 56°N, 175°E essentially removed from the fishery most of the area where North American pink salmon are known to co-occur with Asian pink salmon. As seen in Figure 12, Alaskan pink salmon occur in the eastern sector of the present mother-ship fishery area in the Bering Sea, but recent even-year catches in subarea 10 have averaged only 82,000 fish, and the proportion of Alaskan fish within the catch would likely have been small. In summary, qualitative analysis of tagging data suggests that interceptions of North American pinks by the high seas fisheries were probably never substantial (except possibly the 1974 catch of 253,000 fish, estimated by Fredin et al. [1977]).

Coho salmon

Information on origins of coho salmon migrating in the area of the pre-1977 mother-ship and land-based fisheries comes mainly from INPFC-related high seas tagging experiments, and from scale pattern studies by Myers et al. (1981), Walker and Harris (1982), Walker and Davis (1983), Kato (1984), and Kato and Ishida (1985, 1986). The numbers of Asian and North American coastal tag recoveries from 1956-85 releases in May-July between 160°E and 160°W are shown in Figure 13. The Asian recoveries have been from Kamchatka and the northern coast of the Okhotsk Sea. The North American recoveries have nearly all been from western Alaska, but one recovery from releases in the northern part of subarea 9 was recorded from Kodiak Island in central Alaska. In addition, a recent recovery of a coded-wire tagged Oregon coho (not shown in Figure 13) at 53°12'N,

166°52'W (off the southern side of Unalaska Island) in late April signifies that other North American stocks are distributed at least as far west as the eastern Aleutian Islands (Dahlberg et al. 1986). The few tag recoveries available suggest that Asian stocks predominate throughout the pre-1977 mothership and land-based fishery areas, and that Alaskan stocks range only into the eastern sectors which were closed by the 1978 North Pacific Treaty Annex. I consider this conclusion highly provisional, however, due to paucity of data. Certainly, there are too few tag recoveries available to permit the conclusion that North American coho definitely do not range west of 175°E south of the Aleutians or into the central Bering Sea northeast of 56°N, 175°E. A major difficulty in interpreting tagging data for coho in particular is that the inshore exploitation rates (offering opportunity for coastal tag recovery) are not known for many Asian and Alaskan coho stocks.

Myers et al. (1981), Walker and Harris (1982), and Walker and Davis (1983) applied scale pattern analysis to estimate stock composition of age 2.1 coho (the predominant age group in the western north Pacific), according to four categories: Asia,

western Alaska, central Alaska, and southeast Alaska/British Columbia. The high seas samples were collected mainly in June and July, south of 50°N, between 160°E and 175°W. Results for all three sample years 1979-81 indicated that coho from southeast Alaska/British Columbia are either absent or in very low relative abundance in the study area. Stock proportion estimates for western and central Alaskan coho in certain times and areas, however, were often high and statistically significant, and indicated that these stock groups occur farther west and southwest than is known by tagging results. The estimated proportions of western and central Alaskan coho progressively increased in the analyses of 1979, 1980, and 1981 samples; Asian stocks were generally shown to predominate in 1979, but Alaskan stocks predominated in the 1981 study. Walker and Davis (1983) suggested that the 1980 and 1981 studies may have been especially susceptible to statistical biases stemming from scale quality. Japanese scientists subsequently designed studies to elucidate possible biases in coho scale pattern studies that employ U.S.S.R. coastal samples as Asian standards. At the time of this writing, the INPFC nations are working cooperatively

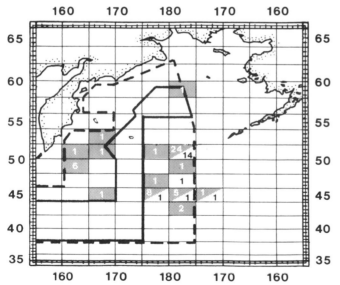

Figure 13. Known ranges of Asian[1] and North American[2] coho salmon between 160°E and 160°W, in May-July.

[1] Partially or fully shaded areas.

[2] Partially or fully open areas containing data.

Notes: Data are the numbers of Asian and/or North American coastal recoveries from tag releases in the areas (1956-85 INPFC-related tagging). The dotted and solid bold lines outline the pre-1977 and 1978-85 high seas salmon fishery areas, respectively.

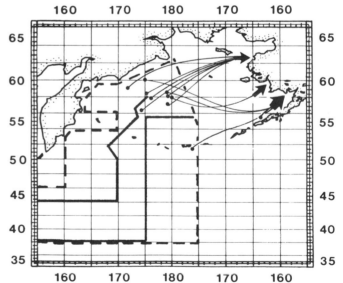

Figure 14. Release and recovery locations for 11 chinook salmon tagged and released during May-July 1956-85 INPFC-related high seas tagging operations between 160°E and 160°W, and recovered in coastal areas.

Note: The dotted and solid bold lines outline the pre-1977 and 1978-85 high seas salmon fishery areas, respectively.

to reduce biases in these studies, and are beginning another series of scale pattern studies of coho (and other species) that may yield more reliable results.

Until the new scale pattern studies resolve the question, I provisionally conclude that major intermingling of Asian and Alaskan coho in the western north Pacific occurs primarily in areas which were excluded from the high seas fisheries by the 1978 North Pacific Treaty Annex. High seas interceptions of North American coho may have been considerable before 1978, especially by the land-based driftnet fishery in subareas 14 and 15, and in some years by the mother-ship fishery in subareas 7 and 9 (see Figure 8). If Alaskan coho indeed migrate west of 175°E, as suggested by scale pattern studies, then there is still opportunity for significant interception of Alaskan coho in subareas 5 and 13, although the present level of potential interceptions would be much less than before 1978.

Chinook salmon

Determination of regional origins of chinook salmon in the western north Pacific Ocean and Bering Sea is hampered by the low abundance of the species in research vessel catches. Virtually all of the information that can be used to infer origins of chinook in the high sea fishery area has come from INPFC-related high seas tagging and from scale pattern analyses by Major et al. (1975, 1977a, b) and Myers et al. (1984).

There have been only eleven coastal recoveries from high seas tag releases in May-July 1956-85, 160°E to 160°W, and all of them have been from western Alaska or the Canadian Yukon (Figure 14). There have been three other notable recoveries. A Kamchatka River recovery from a release in mid-August at 49°35′N, 172°03′W suggests that Asian chinook range across the entire pre-1978 high seas fishery area; and an upper Columbia River recovery of a chinook released in mid-August at 51°29′N, 176°34′W and a southeast Alaskan troll fishery recovery of a chinook tagged at 51°18′N, 178°28′W attest that chinook from North American areas other than western Alaska and Canadian Yukon occur in at least the eastern sectors of the pre-1978 high seas fishery area.

The lack of coded-wire tag recoveries from chinook caught by the mother-ship fishery provides circumstantial information on their origins. In 1981-86, U.S. observers on Japanese salmon mother ships examined nearly 53,000 chinook salmon caught in the U.S. 200-mile zone, and found only seventeen fish missing the adipose fin; none of these carried a coded-wire tag (Dahlberg et al. 1986). Because the great majority of coded-wire tagging of chinook salmon is done in areas south of central Alaska, the lack of coded-wire recoveries from mother-ship catches suggests that chinook from southeast Alaska, British Columbia, and the U.S. Pacific Northwest are in low relative abundance west of 175°E.

Major et al. (1975, 1977a, b) performed discriminant analyses of scale samples collected in the area of the mother-ship fishery. Their model had two stock-group categories: Asia and western Alaska (including Canadian Yukon). Their results indicated that western Alaskan fish predominated in the Bering Sea and in subareas 7 and 9 south of the Aleutians, and that Asian fish predominated in subareas 1, 3 and 5. Myers et al. (1984) performed scale pattern analyses of 1975-81 samples from the mother-ship and land-based areas, according to four categories: Asia, western Alaska (including Canadian Yukon), central Alaska, and southeast Alaska/British Columbia. Stock proportion estimates calculated by Myers et al. (1984) and summarized by Myers et al. (in press) indicated that western Alaskan chinook greatly predominated in the Bering Sea. For the region south of the Aleutian Islands, however, the scale analysis showed that central Alaskan fish generally predominated, that Asian and western Alaskan fish were in lower proportion, and that southeast Alaskan and British Columbian fish were in very low relative abundance. Several follow-up studies were conducted by Japan Fisheries Agency and Fisheries Research Institute scientists to examine possible sources of statistical bias in scale analyses (Knudsen and Davis 1985, Ito et al. 1985, 1986, and Myers 1985, 1986). At the time of this writing, the INPFC nations are beginning a cooperative program to standardize scale pattern analysis methodology, and to perform a new series of analyses to determine origins of chinook salmon and other species in the high seas fishery area.

Based on information currently available, I conclude that most of the chinook salmon caught by the mother-ship fishery and perhaps also by the land-based fishery are of North American origin, principally from western Alaska, the Canadian Yukon, and central Alaska. Tagging results and two independent scale pattern studies indicated that most chinook in the central Bering Sea are of western Alaskan and Canadian Yukon origin. Chinook migrating south of the Aleutian Islands are of more diverse regional origin.

Myers et al. (1984) used the results of their scale pattern study to estimate the high seas chinook catches by regional stock group. The estimated interceptions of North American chinook by the mother-ship fishery declined from a mean of 180,000 fish in 1972-77 to a mean of 81,000 fish in 1978-83 (not including the anomalously high 1980 value). This decrease in chinook interceptions was due mostly to the overall decrease in total catch. The exceptionally large 1980 catch would not likely occur again, even in a year of high abundance, because (1) there are presently quotas on high seas chinook catch (210,000 and 162,000 fish in 1985 and 1986, respectively) whereas there were no chinook quotas before 1983; and (2) the 1986 North Pacific Treaty Annex will phase out mother-ship operations in the central Bering Sea, where chinook tend to be in highest abundance. High seas interceptions of western Alaskan and Canadian Yukon chinook are expected to decrease after 1994 when the mother-ship fishery ceases fishing north of 56°N in the Bering Sea, but some catches of this stock group will continue in subareas 3, 5 and 6. Myers et al. (1984) noted that their interception estimates for the land-based fishery were provisional because of the paucity of scale pattern results for the area south of 46°N. There is presently insufficient information to determine what effect closure of the area east of 175°E (or 174°E) had on stock composition of land-based fishery chinook catches.

Steelhead trout

"Steelhead," as used in this paper, includes the North American anadromous trout *Salmo gairdneri* and the nearly indistinguishable anadromous Kamchatkan trout *Salmo mykiss* (Behnke 1966, Savvaitova et al. 1973, Sutherland 1973, Maksimov 1976, and Okazaki 1984). Steelhead were not emphasized for many years in INPFC-related high seas research, but the 1978 North Pacific Treaty revision placed greater importance on them, and a considerable amount of research has been done

since 1978 to determine abundance, age and maturity composition, distributions of regional-stock and racial groups, food habits, and other aspects of the species's ocean life history. Several of these studies are still in progress.

The general ocean distribution of steelhead was described by Sutherland (1973) and Okazaki (1983). Steelhead tend not to range far into the Bering Sea, but they occur across the entire north Pacific south of the Aleutian Islands and throughout the Gulf of Alaska. Abundance of steelhead is very low in the far western Pacific (near the Kuril Islands and south of Kamchatka), and increases eastward. As is the case for some of the Pacific salmons (particularly immature fish), the overall distribution of steelhead tends to shift northward during the summer months, such that the latitudinal center of distribution (in waters west of 175°W) shifts from about 44°N in May to about 49°N in August (rough interpretation of charts in Okazaki 1983).

Figure 15 shows the distribution of North American steelhead in the north Pacific (west of 160°W), as known from 1956-85 high seas tagging in May-July, from the recovery of coded-wire tagged fish

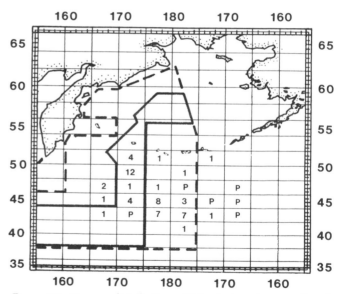

Figure 15. Known distribution of North American steelhead trout in waters between 160°E and 160°W, as indicated by 1956-85 INPFC-related studies.

Notes: Data are the numbers of occurrences of North American steelhead in 2°-latitude by 5°-longitude areas in May-July, as known by coastal recoveries of high seas tags or by high seas recovery of coded-wire tags. "P" indicates that there are no tagging data, but that U.S. Pacific Northwest fish are known to occur in the area by detection of origin-indicating parasites (see text).

through 1986, and from the detection of two parasites that indicate U.S. Pacific Northwest origin (Margolis 1984, 1985). North American steelhead range across the present mother-ship fishery area and across some of the most heavily fished sectors of the land-based fishery area, as far west as about 167°E. Recovery of tags indicates that North American steelhead distributed west of 175°E represent stocks from the U.S. Pacific Northwest, British Columbia, and southeast Alaska. The oceanic distribution of the Asian form is poorly known, and can be inferred from only one study by Okazaki (1985). That study, based on gene frequency differences between the Asian form and the North American coastal and inland forms, suggested that the Asian form is in high relative abundance in the western and central north Pacific and even in the central Gulf of Alaska. This finding was qualified by the author, and is indeed doubtful because information in Savvaitova et al. (1973) and from the U.S.S.R. fishery research agency TINRO (personal communication) suggests that the Asian population consists of small runs to a few river systems, and is smaller than the combined North American population by orders of magnitude. I provisionally conclude that the great majority of steelhead migrating in the Japanese high seas salmon fishery area east of about 165°E are of North American origin.

Assessing the magnitudes of catch of North American steelhead by the Japanese high seas salmon fisheries is difficult, because the mother-ship fishery does not provide steelhead catch statistics, and the land-based fishery began to provide steelhead statistics only in 1981. Steelhead are not retained or processed by the mother-ship fishery, and incidental catches of the species are not recorded. Scientists at the Fisheries Research Institute are presently estimating incidental steelhead catches by the mother-ship fishery, by applying steelhead catch-per-unit-effort indices calculated from Japanese research vessel data to mother-ship effort statistics. An earlier effort to make such estimates suggested that the mother-ship fishery incidentally caught an average of about 4,600 fish/year (range 800 to 13,300 fish) in 1978-83 (Fisheries Research Institute, unpublished data) but these estimates may be revised following an updated analysis. Japan reported that the land-based fishery made catches of about 23,000, 27,000, 29,000, 15,000, 14,000, and 8,000 fish in the years 1981 through 1986 (the 1981 figure

is extrapolated from statistics reported by about 67 percent of the fleet).

Although firm conclusions regarding the high seas fishery's catches of North American steelhead must await completion of ongoing studies, some qualitative inferences can be made. Abundance of steelhead in subareas 7, 9, 14 and 15 is generally higher than in the subareas just west of 175°E (Okazaki 1983), so the closure of the area east of 175°E in 1978 would certainly have served well to protect North American steelhead from high seas capture. However, there is potential for increased steelhead catches in areas remaining open to the mother-ship and land-based fisheries, which would partially offset savings of North American steelhead gained from closure east of 175°E. Because of their summertime northward migrations, steelhead are most vulnerable to catch by the mother-ship fishery in July. In July 1972-77 fishing effort in subareas 3, 5, 7 and 9 averaged 621,000 tans, whereas in July 1978-85 effort in subareas 3 and 5 averaged 988,000 tans. Thus, mother-ship catches of North American steelhead west of 175°E undoubtedly increased. The southward shift of land-based fishing effort west of 175°E forced by the northern boundary at 44°N (except 46°N in subarea 13 before mid-June) may also have led to an increase in steelhead catches west of 175°E. Further analysis is required to determine the net changes in mother-ship and land-based catches of North American steelhead after 1977.

SUMMARY

A change in international fisheries policy led many nations in the mid- and late 1970s to claim 200-mile fishery conservation zones, and also led Pacific Rim nations with major runs of anadromous salmonids to claim increased management authority over their salmonid stocks on the high seas. This change toward extended jurisdiction of fishery resources prompted revisions in 1977-78 of the North Pacific Treaty, which was signed in 1952 between Canada, Japan, and the United States, and the U.S.S.R.-Japan Treaty, signed in 1956. These treaty revisions included new protocols that greatly changed the times and areas of operation of the Japanese mother-ship and land-based driftnet fisheries for salmon and imposed a much more restrictive quota system than before 1977. Sharp reductions in catches have resulted, as well as major changes in the composition of the catches. For instance, the proportion of sockeye salmon in the mother-ship catch has doubled since 1978 because of increased catches of immature sockeye and because the pink salmon catch greatly decreased with the elimination of certain western sectors. The land-based fishery's catch of sockeye salmon, however, greatly decreased after 1977.

Assessment of catches of North American fish for each species is difficult because information on origins of anadromous salmonids in the high seas fishery areas varies greatly in nature and quantity between species, and in many cases is not conducive to quantitative use. Results from several past studies were used to make quantitative estimates of interceptions of North American sockeye salmon by the high seas fisheries. The principal North American stock-group caught by these fisheries is from western Alaska (particularly Bristol Bay); some central Alaskan and southeast Alaskan/British Columbian sockeye were intercepted in the now-closed eastern sectors before 1978, but interceptions of these stock-groups have probably remained negligible since 1978. Total interceptions of western Alaskan sockeye by the mother-ship fishery decreased since 1978, but while catches of maturing fish decreased (due largely to restriction of early-season fishing), catches of immatures actually increased (due to heavy fishing south of the Aleutians in July). The proportion of fish of Alaskan origin in the mother-ship sockeye catches has declined since 1977, but not as much as might be expected in view of the closure southeast of 56°N, 175°E because of offsetting effects from closure northwest of 44°N, 170°E. The elimination of the sector southeast of 56°N, 175°E certainly served to reduce potential interceptions of North American sockeye, but actual savings of North American sockeye in recent years are due also in part to the lowered catch quotas imposed by the U.S.S.R.-Japan protocols. Current research is directed at deriving quantitative estimates of interception of North American fish for the other Pacific salmon species and for steelhead trout.

ACKNOWLEDGMENTS

Contribution no. 718, School of Fisheries, WH-10, University of Washington, Seattle, WA 98195. This work was done under NOAA research contract no. 85-ABC-00006.

LITERATURE CITED

Atkinson, C.E. 1964. The salmon fisheries of the Soviet Far East. M.S. Thesis, Univ. Washington, Seattle. 137 pp.

Behnke, R.J. 1966. Relationships of the Far Eastern trout, *Salmo mykiss* Walbaum. *Copeia* 1966:346-348.

Chitwood, P.E. 1969. Japanese, Soviet, and South Korean fisheries off Alaska; Development and history through 1966. U.S. Dept. Int., Bur. Comm. Fish., Circ. 310. 34 pp.

Cook, R.C., K.W. Myers, R.V. Walker and C.K. Harris. 1981. The mixing proportion of Asian and Alaskan sockeye salmon in and around the landbased driftnet fishery area, 1972-1976. (Document submitted to annual meeting of Int. N. Pac. Fish. Comm., 1981). Fisheries Research Inst., Univ. Washington, Seattle. 81 pp.

Dahlberg, M.L., F.P. Thrower and S. Fowler. 1986. Incidence of coded-wire-tagged salmonids in catches of foreign commercial and research vessels operating in the North Pacific Ocean and Bering Sea in 1985-1986. (Document submitted to annual meeting of Int. N. Pac. Fish. Comm., 1986). Auke Bay Laboratory, Northwest and Alaska Fish. Centr., Natl. Mar. Fish. Serv., Auke Bay, Alaska. 26 pp.

Fredin, R.A. and D.D. Worlund. 1974. Catches of sockeye salmon of Bristol Bay origin by the Japanese mothership salmon fishery, 1956-70. *Int. N. Pac. Fish. Comm., Bull.* 30:1-80.

Fredin, R.A., R.L. Major, R.G. Bakkala and G.K. Tanonaka. 1977. Pacific salmon and the high seas salmon fisheries of Japan. Unpubl. Processed Rep., U.S. Natl. Mar. Fish. Serv., Northwest and Alaska Fish. Centr., Seattle. 324 pp.

French, R.R., H. Bilton, M. Osako and A. Hartt. 1976. Distribution and origin of sockeye salmon (*Oncorhynchus nerka*) in offshore waters of the North Pacific Ocean. *Int. N. Pac. Fish. Comm., Bull.* 34. 113 pp.

Fukuhara, F.M. 1955. Japanese high-seas mothership-type drift gill-net salmon fishery—1954. *Comm. Fish. Rev.* 17(3):1-12.

Fukuhara, F.M. 1971. An analysis of the biological and catch statistics of the Japanese mothership salmon fishery. Ph.D. Dissertation, Univ. Washington, Seattle. 305 pp.

Godfrey, H., K.A. Henry and S. Machidori. 1975. Distribution and abundance of coho salmon in offshore waters of the North Pacific Ocean. *Int. N. Pac. Fish. Comm., Bull.* 31:1-80.

Gregory, H.E. and K. Barnes. 1939. *North Pacific fisheries with Special Reference to Alaska Salmon.* Studies of the Pacific, No. 3, American Council, Inst. of Pacific Relations, Inc. 322 pp.

Harris, C.K. 1978. Patterns of fishing effort and catch by the Japanese landbased driftnet fishery. Unpubl. Processed Rep., Fisheries Research Inst., Univ. Washington, Seattle. 21 pp.

Harris, C.K. (1987) Catches of North American sockeye salmon (*Oncorhynchus nerka*) by the Japanese high seas salmon fisheries, 1972-84. Pages 458-479 in: Smith, H.D., L. Margolis and C. Wood, eds. *Sockeye salmon (Oncorhynchus nerka) Population Biology and Future Management.* Can. Spec. Publ. Fish. Aquat. Sci., No. 96.

Herrfurth, A.G. 1985. Japan's Pacific salmon fisheries and trade, 1974-84. *Mar. Fish. Rev.*, 47(3):78-82.

Hollick, A.L. 1981. *U.S. Foreign Policy and the Law of the Sea.* Princeton Univ. Press, Princeton. 496 pp.

Ignell, S.E. and M.L. Dahlberg. 1958. An analysis of the distribution of fishing effort and catch per unit effort in the Japanese landbased driftnet fishery, 1978-1984. (Document submitted to annual meeting of Int. N. Pac. Fish. Comm., 1985). Auke Bay Laboratory, Natl. Mar. Fish. Serv., Auke Bay, Alaska. 24 pp.

International North Pacific Fisheries Commission. 1979. Historical catch statistics for salmon of the North Pacific Ocean. *Int. N. Pac. Fish. Comm., Bull.* 39:1-166.

Ishida, Y., S. Ito and K. Takagi. 1983. An analysis of scale patterns of Japanese hatchery-reared chum salmon in the North Pacific Ocean. (Document submitted to annual meeting of Int. N. Pac. Fish. Comm., 1983). Japan Fisheries Agency, Tokyo, Japan. 24 pp.

Ishida, Y., S. Ito and K. Takagi. 1984. Further analysis of scale patterns of Japanese hatchery-reared chum salmon in the North Pacific Ocean. (Document submitted to annual meeting of Int. N. Pac. Fish. Comm., 1984). Japan Fisheries Agency, Tokyo, Japan. 22 pp.

Ishida, Y., S. Ito and K. Takagi. 1985. Stock identification of chum salmon based on scale pattern by discriminant function. (Document submitted to annual meeting of Int. N. Pac. Fish. Comm., 1985). Japan Fisheries Agency, Tokyo, Japan. 15 pp.

Ito, J., Y. Ishida, and S. Ito. 1985. Stock identification of chinook salmon in offshore waters in 1974 based on scale pattern analysis. (Document submitted to annual meeting of Int. N. Pac. Fish. Comm., 1985.) Japan Fisheries Agency, Tokyo, Japan. 18 pp.

Ito, J., Y. Ishida, and S. Ito. 1986. Further analysis of stock identification of chinook salmon in offshore waters in 1974. (Document submitted to annual meeting of Int. N. Pac. Fish. Comm., 1986.) Japan Fisheries Agency, Tokyo, Japan. 19 pp.

Jackson, R.I. and W.F. Royce. 1986. *Ocean Forum; An Interpretative History of the International North Pacific Fisheries Commission.* Fishing News Books, Ltd., Farnham. 240 pp.

Johnson, R.W. 1967. The Japan-United States salmon conflict. *Wash. Law Rev.* 43(1):1-43.

Kasahara, H. and W. Burke. 1973. North Pacific fisheries management. Program of International Studies of Fisheries Assessments. Resources for the Future, Inc., Washington, D.C. 91 pp.

Kato, M. 1984. Preliminary examination related to scale pattern analysis of coho salmon. (Document submitted to annual meeting of Int. N. Pac. Fish. Comm., 1984). Japan Fisheries Agency, Tokyo, Japan. 23 pp.

Kato, M. and Y. Ishida. 1985. Scale pattern analysis of coho salmon in the northwest North Pacific Ocean by materials obtained in 1975. (Document submitted to annual meeting of Int. N. Pac. Fish. Comm., 1985). Japan Fisheries Agency, Tokyo, Japan. 18 pp.

Kato, M. and Y. Ishida. 1986. Scale pattern analysis of coho salmon in the northwest North Pacific Ocean using materials obtained by salmon research vessels in 1976. (Document submitted to annual meeting of Int. N. Pac. Fish. Comm., 1986). Japan Fisheries Agency, Tokyo, Japan. 17 pp.

Knudsen, C.M. and C.K. Harris. 1982. Occurrence of British Columbia and southeastern Alaska sockeye salmon in and near the Japanese landbased driftnet fishery area, 1972-1976. (Document submitted to annual meeting of Int. N. Pac. Fish. Comm., 1982). Fisheries Research Inst., Univ. Washington, Seattle. 38 pp.

Knudsen, C.M. and N.D. Davis. 1985. Variation in salmon scale characters due to body area sampled. (Document submitted to annual meeting of Int. N. Pac. Fish. Comm., 1985). Fisheries Research Inst., Univ. Washington, Seattle. FRI-UW-8504. 59 pp.

Leonard, L. 1944. *International Regulation of Fisheries.* Carnegie Endowment for International Peace, Div. Internatl. Law, Monogr. No. 7. Washington, D.C. 201 pp.

Major, R.L., S. Murai and J. Lyons. 1975. Scale studies to identify Asian and Western Alaskan chinook salmon. Pages 80-97 in Int. N. Pac. Fish. Comm., Annual Rep. 1973.

Major, R.L., S. Murai and J. Lyons. 1977a. Scale studies to identify Asian and Western Alaskan chinook salmon: the 1969 and 1970 Japanese mothership samples. Pages 78-81 in Int. N. Pac. Fish. Comm., Annual Rep. 1974.

Major, R.L., S. Murai and J. Lyons. 1977b. Scale studies to identify Asian and Western Alaskan chinook salmon. Pages 68-71 in Int. N. Pac. Fish. Comm., Annual Rep. 1975.

Major, R.L., J. Ito, S. Ito and H. Godfrey. 1978. Distribution and origin of chinook salmon (*Oncorhynchus tshawytshcha*) in offshore waters of the North Pacific Ocean. *Int. N. Pac. Fish. Comm., Bull.* 38:1-54.

Maksimov, V.A. 1976. Parallelnaya izmenchivost u vidov lososei roda *Salmo*. *Vopr. Ikhtiologii* 16,5(100):765-772. (Transl.: Parallel variability in species of the genus *Salmo*. *J. Ichthyology* 16(5):693-701.)

Manzer, J.I., T. Ishida, A.E. Peterson and M.G. Hanavan. 1965. Salmon of the North Pacific Ocean—Part V: Offshore distribution of salmon. *Int. N. Pac. Fish. Comm., Bull.* 15:1-452.

Margolis, L. 1963. Parasites as indicators of the geographical origin of sockeye salmon, *Oncorhynchus nerka* (Walbaum), occurring in the North Pacific Ocean and adjacent waters. *Int. N. Pac. Fish. Comm., Bull.* 11:101-156.

Margolis, L. 1984. Preliminary report on identification of origin of ocean-caught steelhead trout, *Salmo gairdneri*, using naturally occurring parasite "tags." Dept. Fisheries and Oceans, Fish. Res. Branch, Pac. Biol. Sta., Nanaimo, B.C., Canada (Unpubl. report). 23 pp.

Margolis, L. 1985. Continent of origin of steelhead, *Salmo gairdneri*, taken in the North Pacific Ocean in 1984, as determined by naturally occurring parasite "tags." (Document submitted to annual meeting of Int. N. Pac. Fish. Comm., Tokyo, Japan, November 1985.) Dept. Fisheries and Oceans, Fish. Res. Branch, Pac. Biol. Sta., Nanaimo, B.C., Canada. 18 pp.

Meyer, W.G. and C.K. Harris. 1983. Interceptions of North American and Bristol Bay sockeye salmon by the Japanese landbased driftnet fishery, 1972-1981. (Document submitted to annual meeting of the Int. N. Pac. Fish. Comm., 1983). Fisheries Research Inst., Univ. Washington, Seattle. 58 pp.

Moiseev, P.A. 1956. Promysel lososei v otkrytom more v severnoi chasti Tikhogo Okeana. *Ryb. Khoz.* 32(4):54-59. (Transl.: High seas salmon fisheries in the North Pacific. *In:* 1961. *Pacific salmon—selected articles from Soviet periodicals.* Israel Progr. for Sci. Transl., in coop. with U.S. Natl. Sci. Found. and Dept. Int.)

Myers, K.W. 1985. Racial trends in chinook salmon (*Oncorhynchus tshawytscha*) scale patterns. (Document submitted to annual meeting of Int. N. Pac. Fish. Comm., 1985.) Univ. Washington, Seattle. Fisheries Research Inst., FRI-UW-8503. 56 pp.

Myers, K.W. 1986. The effect of altering proportions of Asian chinook stocks on regional scale pattern analysis. (Document submitted to annual meeting of Int. N. Pac. Fish. Comm., 1986.) Fisheries Research Inst., Univ. Washington, Seattle. FRI-UW-8605. 44 pp.

Myers, K., R. Cook, R. Walker and C. Harris. 1981. The continent of origin of coho salmon in the Japanese landbased driftnet fishery area in 1979. (Document submitted to annual meeting of Int. N. Pac. Fish. Comm., 1981.) Fisheries Research Inst., Univ. Washington. Seattle. 34 pp.

Myers, K.W., D.E. Rogers, C.K. Harris, C.M. Knudsen, R.V. Walker, and N.D. Davis. 1984. Origins of chinook salmon in the area of the Japanese mothership and landbased driftnet salmon fisheries in 1975-1981. (Document submitted to annual meeting of Int. N. Pac. Fish. Comm., 1984.) Fisheries Research Inst., Univ. Washington, Seattle. 208 pp.

Myers, K.W., C.K. Harris, C.M. Knudsen, R.V. Walker, N.D. Davis, and D.E. Rogers. (In press) Stock origins of chinook salmon in the area of the Japanese mother-ship salmon fishery. *N. Amer. J. Fish. Management.*

Nagasaki, F. 1967. Some Japanese far-sea fisheries. *Wash. Law Rev.*, 43(1):197-229.

Neave, F., T. Yonemori and R.G. Bakkala. 1976. Distribution and origin of chum salmon in offshore waters of the North Pacific Ocean. *Int. N. Pac. Fish. Comm., Bull.* 35:1-79.

Neo, M. 1964. How the salmon fleets operate. Pages 428-431 in *Modern Fishing Gear of the World.* Vol. 2. Fishing News (Books), Ltd., London.

Okazaki, T. 1983. Distribution and seasonal abundance of *Salmo gairdneri* and *Salmo mykiss* in the North Pacific Ocean. *Jap. J. Ichthyology* 30(3):235-246.

Okazaki, T. 1984. Genetic divergence and its zoogeographical implications in closely related species *Salmo gairdneri* and *Salmo mykiss*. *Jap. J. Ichthyology* 31(3):297-311.

Okazaki, T. 1985. Distribution and migration of *Salmo gairdneri* and *Salmo mykiss* in the North Pacific based on allelic variations of enzymes. *Jap. J. Ichthyology* 32(2):203-215.

Parker, W.B. 1974. Alaska and the law of the sea; International fisheries regimes of the North Pacific. Publ. by Arctic Environment Inform. and Data Centr., Alaska Sea Grant Rep. No. 73-13. 65 pp.

Rogers, D.E. 1984. Trends in abundance of northeastern Pacific stocks of salmon. Pages 100-127 in Pearcy, W.G., ed. *The influence of ocean conditions on the production of salmonids in the North Pacific—A workshop.* Coop. Inst. Mar. Resource Studies, Oregon State Univ. Sea Grant College Program, Corvallis. (ORESU-W-83-001.)

Savvaitova, K.A., V.A. Maksimov, M.V. Mina, G.G. Novikov, L.V. Kokhmenko and V.Ye. Matsuk. 1973. *Kamchatskie blagorodnye lososi (Kamchatka trout).* Voronezh Univ. Press.

Sutherland, D.F. 1973. Distribution, seasonal abundance, and some biological features of steelhead trout, *Salmo gairdneri*, in the North Pacific Ocean. *Fish. Bull.* 71(3):787-826.

Taguchi, K. 1966. *The Salmon Resources and Salmon Fisheries around the Pacific Ocean.* Koseisha Koseikaku. Tokyo. 390 pp. (Prelim. partial transl. of chapters 6 and 7 by U.S. Joint Publ. Research Serv. for U.S. Bur. Comm. Fish., Seattle).

Takagi, K., K.V. Aro, A.C. Hartt and M.B. Dell. 1981. Distribution and origin of pink salmon (*Oncorhynchus gorbuscha*) in offshore waters of the North Pacific Ocean. *Int. N. Pac. Fish. Comm., Bull.* 40:1-195.

VanCleve, R. and R.W. Johnson. 1963. Management of the high seas fisheries of the northeastern Pacific. Contr. No. 160, Coll. of Fisheries, Univ. Washington, Publ. in Fish., New Series, Vol. II(2). 63 pp.

Walker, R. and C. Harris. 1982. The continent of origin of coho salmon in the Japanese landbased driftnet fishery area in 1980. (Document submitted to annual meeting of Int. N. Pac. Fish. Comm., 1982). Fisheries Research Inst., Univ. Washington, Seattle. 26 pp.

Walker, R. and N. Davis. 1983. The continent of origin of coho salmon in the Japanese landbased driftnet fishery area in 1981. (Document submitted to annual meeting of Int. N. Pac. Fish. Comm., 1983). Fisheries Research Inst., Univ. Washington, Seattle. 48 pp.

Wertheim, B. 1935. The Russo-Japanese fisheries controversy. *Pac. Affairs* 8(2):185-198.

Factors Affecting Survival of Coho Salmon off Oregon and Washington

William G. Pearcy

Professor of Oceanography
Oregon State University

The purpose of this paper is to emphasize the need to understand the effects of the ocean environment on the survival of salmonid stocks. We can no longer afford to view the north Pacific Ocean as an invariate, unlimited, and benign pasture for increasing releases of salmon. Better knowledge of ocean conditions and how they affect distribution, migrations, and survival of salmon will be needed to allow more effective policies of management and enhancement in the future.

This paper focuses on coho salmon produced in the Oregon Production Index (OPI) area which includes the Columbia River, Oregon, and California regions. OPI coho stocks are some of the most thoroughly studied in the world. The history of releases of hatchery smolts and the production (catch plus escapement) of adult coho in the OPI area are shown in Figure 1 (Oregon Department of Fish and Wildlife 1985; Nickelson 1986). Four phases are recognized:

— The 1960s, when large-scale hatchery production of smolts was initiated to circumvent freshwater limitations to production, and a gratifying positive relationship was found between smolts released and adult production.

— 1970-75, a period when smolt releases from hatcheries were fairly constant (about 35 million) but adult production varied dramatically.

— 1976-81, a period of poor adult production and survival which coincided with increased releases of hatchery smolts. The increased smolt production during this period was largely from private hatcheries; their releases peaked at twenty-four million smolts in 1981, giving a total hatchery release of about sixty million.

— 1982-85, when the numbers of smolts released declined (due to reduced production by private hatcheries) and adult production, after reaching historic lows of less than one million in 1983 and 1984, increased to about two million adults for the smolt migration year of 1985.

An inverse relationship has generally existed between smolts and adult production since 1975. Two general explanations have been explored to explain the poor production of adult coho in the ocean since the 1975 smolt migration year (Walters in this volume mentions other possible explanations):

— Changes in ocean conditions resulting in adverse survival independent of density; and

— Density-dependent mortality when smolt releases were high.

There is cogent evidence for changing ocean conditions during the last 10 to 20 years. Sea-level and sea-surface temperature increased in the northeastern Pacific between 1970 and 1984, with a large jump in 1976 (McLain 1984, Norton et al. 1985, Royer 1985). These changes were associated with reduced southerly transport of water masses and weak upwelling along the coast.

Marine survival of coho in the OPI is also correlated with the strength of coastal upwelling during the period coho smolts migrate to sea (Gunsolus 1978, Scarnecchia 1981). Nickelson (1986) shows a significant positive relationship between upwelling during March through September at 42°N, 125°W and the survival of hatchery coho during the year of outmigration of smolts. Survival in strong upwelling years was about twice that in weak upwelling years. Interestingly, no

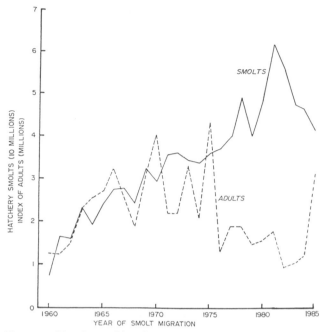

Figure 1. Numbers of hatchery smolts released and estimated abundance (catch and escapement) of hatchery and wild adults in the following year for the Oregon Production Area.
Source: Oregon Department of Fish and Wildlife.

correlation has been observed between survival of wild smolts and upwelling, or for survival of hatchery smolts when considered separately within strong or weak upwelling years. Ocean temperatures in weak upwelling years were also poorly correlated with survival of either hatchery or wild smolts.

The hypothesis of density-dependent survival of coho salmon has been examined by Peterman (1982), Peterman and Routledge (1983), McCarl and Rettig (1983), McGie (1984), and Nickelson (1986). McGie (1984) concluded that upwelling was a significant factor in determining adult production over a 22-year period, but when analyses were made for strong and weak upwelling years separately, adult survival was primarily related to the numbers of smolts released. He suggested that the decline in abundance of coho from 1976 to 1980 was the result of density-dependent mortality caused by the release of large numbers of smolts during a period of weak upwelling.

Nickelson (1986) found no evidence for density-dependent survival for public hatchery, private hatchery, or wild stocks of coho analyzed separately for all years. However, when hatchery and wild populations were combined for all years or for weak upwelling years, the smolt-to-adult relationship was nonlinear, indicating density dependence, as reported by McGie (1984), Peterman and Routledge (1983), and McCarl and Rettig (1983). Nickelson explains this nonlinearity as an artifact due to the shift from primarily wild fish with intrinsically high survival rates in early years to primarily hatchery fish with low survival in recent years. Nickelson concludes that good evidence for density dependence is lacking and that the recent decline in coho production in the OPI area is associated with adverse ocean conditions and is independent of density.

There are three questions that need to be answered if we are to understand the effects of the upwelling and the marine environment on survival of coho salmon in the ocean:

— When in the ocean life history is mortality the highest and most variable?
— Where in the ocean does this critical phase in the life history occur? Is it in estuarine, coastal, or distant oceanic waters?
— What is the mechanism that causes the high mortality and determines year-class success?

In order to learn more about the ocean ecology of juvenile salmon we conducted cruises in the period 1979-85 using fine-meshed purse seines to study their distribution, abundance, movements, food habits, and growth in the ocean off Oregon and Washington. Little was known about early ocean life of juvenile salmon in this region prior to our cruises. We did not even know how long they are present in coastal waters during summer months.

WHEN IS RUN SIZE DETERMINED?

The positive relationship between upwelling during the first spring and summer at sea and subsequent survival of OPI coho, and the close relationship between the number of precocious males (jacks) and the number of adults produced the following year in the OPI area (except for the 1983 El Niño year), indicate that year-class success or run size is determined during the first four months in the ocean, before the time of return of jacks which spend only one summer in the ocean (Gunsolus 1978, Peterman 1982).

Our catches of juvenile salmon in the ocean during June are closely correlated with the OPI

jack index (Figure 2). Mean catch per set of juveniles between the Willapa Bay and the Alsea River and the jack index, 1979-1985, show similar trends, suggesting that variable survival of the year class occurs before or during June. This is only a month after peak outmigration of smolts from the Columbia River, the main source of smolts in the OPI area. Hence the first month in the ocean may be the critical period. This conclusion is supported by our unpublished observations that survival of hatchery coho is more strongly correlated with cumulative upwelling values from March to June than from July through September during the year of smolt migration.

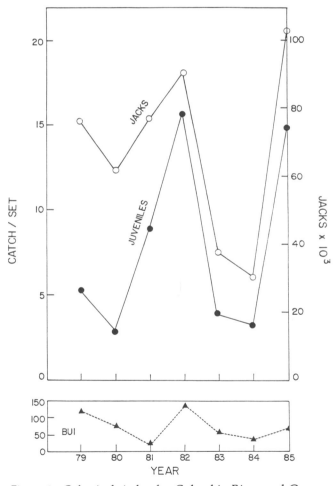

Figure 2. Coho jack index for Columbia River and Oregon-California coastal streams; the average catch per set of juvenile coho salmon caught in purse seines from Willapa Bay, Washington, to the Alsea River, Oregon; and the sum of the monthly mean May and June Bakun upwelling indices (BUI) at 45°N, 125°W, 1979-85.

Although the years 1979 through 1985 were weak upwelling years, as classified by Nickelson (1985), large differences in survival occurred, nevertheless. Smolt survival, as indicated by the jack index, was high in 1982 and 1985 (moderate upwelling) and low in 1983 and 1984 (weak upwelling). These differences in survival may be related to interannual variations in abundance, distribution, and growth of juvenile coho salmon.

WHERE ARE JUVENILE COHO DURING THEIR FIRST SUMMER IN THE OCEAN?

The pioneering studies of Hartt (1980) and Hartt and Dell (1986) provide the most extensive information on distribution and migration of juvenile salmonids in the northeastern Pacific Ocean. Based on the subsequent recovery of coho salmon that were captured and tagged during their first summer in the ocean, these researchers, working in the late 1950s and early 1960s, concluded that coho salmon from the Columbia River, Oregon, Washington, and California comprised a significant proportion of the population of juvenile coho salmon found in waters north of 50°N in the northeast Pacific Ocean. Unfortunately, Hartt and Dell did not sample south of the Strait of Juan de Fuca, so the actual proportion of highly migratory and less migratory coho could not be determined.

Our recent studies off Oregon and Washington also indicate that juvenile coho move to the north during the summer, but these movements may not be extensive and many juvenile coho may reside in coastal waters off Oregon and Washington during the summer rather than migrating north into Alaskan waters. Directional purse seine sets off Oregon and Washington usually caught more juvenile coho in south-facing sets, suggesting northern movement (Miller et al. 1983, Pearcy 1984). Our data on catch per set and length-frequency distributions during the summer also support the conclusion that movement occurs from Oregon to Washington coastal waters during the summer. The catch of juvenile coho salmon in late summer off the Columbia River and Washington was a substantial fraction of the catch in May and June, indicating that many fish remained in local waters throughout most of the summer and did not migrate out of the area into the Gulf of Alaska (Miller et al. 1983).

Recovery of marked fish in our ocean purse seine collections provides the most direct evidence for movements of juvenile coho. Juvenile coho were often found south of their river of ocean entry during May, perhaps a result of advection of water. Later in the summer, however, they were usually captured north of where they had entered the ocean, demonstrating active swimming against the prevailing coastal currents. In late summer of all years, some juvenile coho were found near the mouths of their home streams.

This tendency for juvenile coho salmon to remain in local coastal waters during their first summer in the ocean is supported by data on the recovery of tagged juvenile fish in fisheries from Alaska to California. For nearly all areas, the largest proportion of the stocks recovered in a geographic region had originated in the same region. For example, twenty of twenty-five marked juvenile coho landed in Alaska were from Alaskan hatcheries.

Based on catch per unit of effort, I estimate that the population of juvenile yearling coho in August and September in the area sampled off Oregon and Washington equalled about 5-6 percent of the total number of hatchery and wild fish (excluding subyearling and California fish) entering the OPI area during 1981-85. This must represent a substantial proportion of the total OPI smolt population at that time, assuming that mortality rates decrease with age in the ocean and recognizing that our sampling was restricted to 51-83 percent of the total area from Cape Flattery to Cape Arago, depending on year.

Based on the above evidence, I conclude that most juvenile coho from the OPI area are not highly migratory. Many stay in local coastal waters off Washington and northern Oregon during their first summer of ocean residence. This helps to explain the coupling between coastal upwelling off Oregon and smolt survival. If variable survival occurs soon after ocean entrance, however, coho production could be correlated to upwelling in the spring or early summer even if fish resided in distant waters later in the summer.

The difference between the conclusion of our work and that of Hartt and Dell raises some interesting issues. The studies were conducted in different regions of the north Pacific, and neither study included the entire geographic range of juvenile coho from North America. There are also some differences related to time. Ocean conditions, as previously noted, have changed during the past two decades. The ocean environment is not constant and may have significantly affected the migrations of coho salmon. There is evidence with Fraser River sockeye salmon that ocean conditions can greatly affect ocean migrations (Groot et al. 1984, Hamilton 1985). The migratory behavior of coho stocks in the late 1950s and early 1960s when Hartt and Dell conducted their research may also have been quite different from that observed in our studies due to the composition of stocks. In early years the coho stocks were largely composed of wild fish, whereas in recent years they are mainly hatchery fish. Perhaps wild fish have a greater tendency to migrate far to the north during the first summer of ocean life. In this context it is interesting to note Nickelson's (1986) finding that survival of hatchery but not wild coho in the OPI is significantly correlated with the intensity of coastal upwelling off Oregon. One would expect that fish that migrate out of the coastal waters of Oregon and Washington would be less affected by upwelling in that region.

WHAT IS THE CAUSE OF MORTALITY?

Many hypotheses have been proposed to explain the recent decline in coho production in the OPI area. Two major hypotheses link survival and upwelling. One involves ocean productivity—strong upwelling results in higher production of phytoplankton, zooplankton, and forage for juvenile coho salmon, hence faster growth of smolts and lower size-selective predation rates. The other hypothesis involves changes in distribution or availability of coho smolts to predators, perhaps independent of size or growth—strong upwelling may transport or stimulate smolts to migrate offshore beyond major concentrations of inshore predators, or produce turbid, phytoplankton-rich waters where smolts are less visible to predators (see Scarnecchia 1981, Pearcy 1984, Nickelson 1986).

The productivity-growth hypothesis appears to be the most easily evaluated. Faster growth rates should be found in high survival/strong upwelling years than in low survival/weak upwelling years. Despite large variations in juvenile survival and upwelling among the years 1981-85, our samples indicate only small interannual variations in growth

rate of juvenile coho estimated from mean lengths of yearling fish or from growth rates of marked juvenile coho captured in the ocean during the summer. Growth rates of jacks, however, showed large interannual variations. Growth rates of jacks returning to the Anadromous, Inc., facility at Coos Bay, Oregon, were 21-34 percent lower in the poor upwelling and El Niño year of 1983 than in other years during the period studied (1981-85). Mean lengths of jacks returning to Fall Creek and Siletz hatcheries on the Oregon coast were also lower in 1983 than in other years. Different ocean distributions of jacks and non-jack juvenile coho may explain this discrepancy. In the late summer of 1983, juvenile coho were found in highest numbers far to the north off Cape Flattery, while jacks would probably have been farther south in warmer, less productive waters (Fisher and Pearcy unpublished).

Our work shows that growth of coho during their first summer and fall in the ocean, based on analysis of scale growth patterns between the time of ocean entrance and the winter annulus for adults returning on the Alsea River, Oregon, was not significantly correlated with upwelling during 6 years of strong and weak upwelling between 1961 and 1971. Similarly, Bottom (1985) found no correlation between upwelling during the time of smolt outmigration and first-year growth in the ocean for coho from the Ten Mile Lakes drainage in southern Oregon. In addition, we found no significant differences in body condition or stomach fullness of juvenile coho salmon caught off Oregon and Washington during summers of relatively weak and strong upwelling, 1981-85. Thus available data do not cogently support the hypothesis that food shortage in weak upwelling years leads to slow growth and poor condition.

The alternative hypothesis that upwelling affects the distribution or availability of coho smolts and their susceptibility to predators is more difficult to assess. We rarely found juvenile salmon in the stomachs of large predatory fishes caught in our purse seines. Juvenile salmon occurred only four times in adult cutthroat trout and twice in black rockfish stomachs in over 1600 stomachs of large pelagic nekton examined from our purse seine collections from 1979 to 1984 (Brodeur et al. 1987). Coho smolts have been reported at times in stomachs of black rockfish caught by charter boats. Smolts released from a private hatchery in late summer were common in stomachs of adult coho caught during August 1-3, 1984. No juvenile salmon were found in the stomachs of 290 Pacific whiting collected in nearshore coastal waters off Oregon and Washington in 1982.

There is good evidence for intense predation by sea birds on smolts released from private hatcheries in Oregon during some years. Juvenile coho salmon became a major component of the diet of the common murre off Coos Bay and Yaquina Bay after large releases of coho smolts were initiated. During these years the numbers of murres feeding close to Coos Bay and Yaquina Bay increased dramatically. Some murres were found to have up to twelve coho smolts in their stomachs in 1981, and were estimated to consume coho smolts at a rate of at least 60,000 per day in 1982 (D. Varoujean, University of Oregon, personal communication, Matthews 1983). Juvenile coho salmon were found in murres collected in and offshore of the Columbia River estuary in May-June 1982, and in murres collected in coastal waters from the Columbia River to Coos Bay in 1984, but juvenile salmon were not a major prey at either location (D. Varoujean, personal communication).

Bayer (1986) reported that fish-eating birds aggregated at the mouth of the estuary within hours after salmon smolts were released from the Oregon Aqua-Foods Inc. facility at Yaquina Bay, Oregon. Smolts were especially vulnerable to bird predation soon after release. In the 1983 El Niño year, when ocean productivity and upwelling were low, there were significantly more feeding murres and gulls present during the first day after release than on subsequent days. A significant relationship was not found during 1982 when upwelling was comparatively strong. Possibly alternative prey were more abundant in coastal waters in 1982 than in 1983.

Some of the barging experiments conducted by Oregon Aqua-Foods, where coho smolts were towed offshore in net pens and released, support the hypothesis that onshore predators may drastically affect survival. Survival of smolts in two of five releases in 1983 and 1984 was about four times that of smolts released on-site into Yaquina Bay. Survival of the other three offshore releases was similar to that of on-site releases in these years (R. Severson, Oregon Aqua-Foods, unpublished). Although data are again scanty, the poor smolt survival years of 1983 and 1984 may have been periods

when predation by birds and perhaps other marine predators was severe because of low availability of other prey. This hypothesis suggests that barging would have a less positive effect on survival during good survival years and, indeed, the survival rates of coho smolts released by Oregon Aqua-Foods in offshore waters were not higher than those for on-site releases in 1985 (R. Severson, personal communication).

If offshore distribution of juvenile coho in the coastal zone is related to high survival, we would expect that coho would be found farther offshore in the good survival years of 1982 and 1985 than the poor survival years of 1983 and 1984. The average catch per set at various distances from shore does not support this contention, however. During 1985, we found many juvenile coho far offshore, but the inshore-offshore distributions in 1982, 1983, and 1984 were not markedly different.

The mechanism linking upwelling and survival of juvenile coho is probably complex. Productivity, availability of alternate forage animals, and offshore distribution may all be important, especially in weak upwelling/poor survival years. Knowledge of the major predators of juvenile coho salmon when they enter the ocean and the factors that affect the abundance and availability of their prey are needed to understand the processes affecting mortality of salmon during early ocean life.

LITERATURE CITED

Bayer, R.D. 1986. Seabirds near an Oregon salmon hatchery in 1982 and during the 1983 El Niño. *Fish. Bull.* 84:279-286.

Bottom, D.L. 1985. Research and development of Oregon's coastal salmon stocks: Coho salmon model. Oregon Dept. Fish. Wildl. Ann. Prog. Rep., 11 p.

Brodeur, R.D., H.V. Lorz, and W.G. Pearcy. 1987. Food habits and dietary variability of pelagic nekton off Oregon and Washington, 1979-1984. NOAA Tech. Rep. NMFS 57, 32 p.

Groot, C., L. Margolis, and R. Bailey. 1984. Does the seaward migration of Fraser River sockeye salmon (*Oncorhynchus nerka*) smolts determine the route of return of adults? Pages 283-292 in J.D. McCleave, G.P. Arnold, J.J. Dodson and W.H. Neill eds., *Mechanisms of Migration in Fishes.* Plenum Press, N.Y.

Gunsolus, R.T. 1978. The status of Oregon coho and recommendations for managing the production, harvest, and escapement of wild and hatchery reared stocks. Oreg. Dept. Fish. Wildl. Fish Div. Proc. Rep., 59 p.

Hamilton, K. 1985. A study of the variability of the return migration route of Fraser River sockeye salmon (*Oncorhynchus nerka*). *Can. J. Zool.* 63:1930-1943.

Hartt, A.C. 1980. Juvenile salmonids in the oceanic ecosystem—the critical first summer. Pages 25-27 in W.J. McNeil and D.C. Himsworth, eds. *Salmonid Ecosystems of the North Pacific.* Oregon State University Press, Corvallis.

Hartt, A.C. and M.B. Dell. 1986. Early oceanic migrations and growth of juvenile Pacific salmon and steelhead trout. *Int. North Pac. Fish. Comm., Bull.* 46, 105 pp.

Matthews, D.R. 1983. Feeding ecology of the common murre (*Uria aalge*) off the Oregon coast. M.S. Thesis, University of Oregon, 108 p.

McCarl, B.A. and R.B. Rettig. 1983. Influence of hatchery smolt releases on adult salmon production and its variability. *Can. J. Fish. Aquat. Sci.* 40:1880-1886.

McGie, A.M. 1984. Evidence for density dependence among coho salmon stocks in the Oregon Production Index area. Pages 37-49 in W.G. Pearcy, ed., *The Influence of Ocean Conditions on the Production of Salmonids in the North Pacific.* Oregon State University Sea Grant College Program, Corvallis. (ORESU-W-83-001).

McLain, D.R. 1984. Coastal ocean warming in the northeast Pacific, 1976-83. Pages 61-86 in W.G. Pearcy, ed., *The Influence of Ocean Conditions on the Production of Salmonids in the North Pacific.* Oregon State University Sea Grant College Program, Corvallis. (ORESU-W-83-001).

Miller, D.R., J.G. Williams, and C.W. Sims. 1983. Distribution, abundance and growth of juvenile salmonids off Oregon and Washington, summer 1980. *Fish. Res.* 2:1-17.

Nickelson, T.E. 1986. Influences of upwelling, ocean temperature, and smolt abundance on marine survival of coho salmon (*Oncorhynchus kisutch*) in the Oregon Production Area. *Can. J. Fish. Aquat. Sci.* 43:527-535.

Norton, J., D. McLain, R. Brainard, and D. Husby. 1985. The 1982-83 El Niño event off Baja and Alta California and its effect on ocean climate context. Pages 44-72 in W.S. Wooster and D.L. Fluharty, eds., *El Niño North.* University of Washington Sea Grant Program.

Oregon Department of Fish and Wildlife. 1985. Coho salmon plan status report. Oregon Dept. Fish Wildl., Portland. 21 p.

Pearcy, W.G. 1984. Where do all the coho go? The biology of juvenile coho salmon off the coasts of Oregon and Washington. Pages 50-60 in W.G. Pearcy, ed., *Influence of Ocean Conditions on the Production of Salmonids in the North Pacific.* Oregon State University Sea Grant College Program, Corvallis. (ORESU-W-83-001).

Peterman, R.M. 1982. Nonlinear relation between smolts and adults in Babine Lake sockeye salmon (*Oncorhynchus nerka*) and implications for other salmon populations. *Can. J. Fish. Aquat. Sci.* 39:904-913.

Peterman, R.M. and R.D. Routledge. 1983. Experimental management of Oregon coho salmon (*Oncorhynchus kisutch*): Designing for yield of information. *Can. J. Fish. Aquat. Sci.* 40:1212-1223.

Royer, T.C. 1985. Coastal temperature and salinity anomalies in the northern Gulf of Alaska, 1970-84. Pages 107-115 in W.S. Wooster and D. L. Fluharty, eds., *El Niño North.* University of Washington Sea Grant Program.

Scarnecchia, D.L. 1981. Effects of streamflow and upwelling on yield of wild coho salmon (*Oncorhynchus kisutch*) in Oregon. *Can. J. Fish. Aquat. Sci.* 38:471-475.

Release Strategies for Coho and Chinook Salmon Released into Coos Bay, Oregon

Ron Gowan

Vice President
Anadromous, Inc.
Corvallis, Oregon

INTRODUCTION

Oregon administrative rules for private hatcheries require that juvenile salmon be released into estuaries or into the ocean. Anadromous, Inc., has been releasing tagged coho and chinook salmon into the Coos River estuary since 1978. Since that time over two hundred discrete groups of fish have been tagged and released. Some fish are tagged to meet state requirements; but the majority are tagged to determine the survival to adult of groups of fish with different rearing/release histories. The objective is simple: to improve adult survival. But although statement of the objective is simple, its achievement is not. An infinite number and degree of manipulations of rearing practices can be performed, and distinguishing among those that benefit survival and those that do not is extremely difficult.

Manipulations can be grouped into two general categories with some degree of overlap. The first category comprises cultural practices designed to optimize survival through manipulation of some characteristics of the fish itself, such as health or smoltification. The second general category comprises manipulations designed to improve the fish's response to conditions external to the facility.

Examples of this are varying time of release, using different methods of releasing fish, or releasing fish at different locations. Sometimes it is difficult to place a manipulation into a category; for example, increasing the size of fish at release. It is not clear whether the larger size at release results in a more physically competent smolt or whether a larger size fish is more able to avoid predators and compete for food in the ocean.

Oregon private hatcheries have achieved more success with manipulation designed to improve response to conditions external to rearing facilities than to manipulations of cultural practices (Gowan and McNeil 1983). Other studies have shown that release strategies can influence adult survival. Johnson (1982) in his review of Oregon hatchery release practices for coho salmon did find that in some cases release strategies were able to affect return rates. Bilton (1978) demonstrated the importance of size and time of release for coho released from Canadian facilities as did Hager and Noble (1976) for Columbia River coho. The time of day when fish are released has been suggested by Perry (1983) as affecting adult survival of chinook salmon released into a Canadian estuary. Hansen (1982) demonstrated that Atlantic salmon released into the estuary or offshore survived at a higher rate than smolts released into the river systems. Wagner (1968) evaluated adult survival of steelhead allowed to migrate into the Alsea River at will, compared to fish that were simply crowded out of the raceway. Evenson and Ewing (1984) also evaluated volitional and nonvolitional release of steelhead into the Rogue River. Both studies showed a higher survival for forced release fish.

In the past eight years Anadromous has evaluated several release strategies as they apply to the estuarine environment encountered in Coos Bay, Oregon. Two strategies, size at release and time of release, have been evaluated for coho and chinook salmon over several years. Two other strategies which are concerned with method of onshore release were evaluated for 1 year. Evaluation of offshore release of smolts is an ongoing program.

TIME OF RELEASE

Anadromous has been releasing coded-wire tagged chinook since 1978 and coho since 1980. Coded-wire tag groups have been released in

Table 1. *Adult percentage return of coho released at Coos Bay, Oregon*

Year of release	Month of release						
	Mar	Apr	May	June	Jul	Aug	Sep
1981	.148	.310	1.58	1.84	2.81		
1982				.098	.335		
1983[a]							
1984				.83	3.69	6.39[b]	
1985			8.69	1.64	1.40	9.31	

[a] 1983 data omitted because of El Niño.

[b] In 1984 June and July releases were yearlings; August releases were zero age coho.

several months of each year. Except for 1983, when all releases were devastated by El Niño, a consistent pattern has emerged. Both spring chinook and coho generally survive better when released into the ocean in July or August than at other times, with August the best month. This result is contradictory to that expected for coho from observing smoltification and the outmigration of wild smolts. Oregon coastal coho outmigrate in April and May. However, the result with chinook is consistent with that observed in the Rogue River system where maximum adult survival was reported for spring chinook which outmigrate in September or October (Ewing 1982). The Anadromous chinook stock is derived from Rogue River spring chinook.

In the period 1981 through 1984 coho released later survived at a higher rate than those released in earlier months (Table 1). In 1985, August releases survived at the highest rate, closely followed by May releases, with releases in June and July demonstrating very poor survival. Spring chinook show a similar pattern for release years 1979-84 (Table 2), with August generally the best month,

Table 2. *Adult percentage return of tagged age 2 spring chinook released at Coos Bay, Oregon*

Year of release	Month of release						
	May	June	Jul	Aug	Sep	Oct	Nov
1979			.474	.897		.260	.155
1980				.685	.063	.124	.000
1981		.159	.445	.342	.498	.057	.128
1982					.1470		
1983[a]							
1984				.650	.450		
1985[b]	.58	.27	0	.32			

[a] 1983 data omitted because of El Niño.

[b] All tagged spring chinook released in 1985 were transported offshore.

Table 3. *Adult return of tagged coho released at Coos Bay, Oregon*

Date of release	Size at release (g)	Return (%)	
29 May 1985	37.2	5.95	
29 May 1985	48.4	8.69	.68
27 June 1985	32.7	1.64	
27 June 1985	43.9	3.79	.43
17 July 1985	36,2	1,61	
17 July 1985	49.2	2.53	.63
15 August 1985	40.9	7.39	
15 August 1985	46.7	9.32	.79

Table 5. *Adult return of forced released and self-released groups of coho salmon released into Coos Bay, Oregon, in 1981 and returning in 1982.*

Month released	% Return	
	Self-release[a]	Forced release[b]
April	.12	.32
May	1.24	1.58
June	.95	1.84
July	1.02	2.81
Mean	.83	1.64

[a] 23-27 days at site.
[b] 30 days at site.

closely followed by July and September. 1986 chinook jack returns suggest that in 1985 that pattern changed, with May releases producing the greatest return and survival decreasing for releases in June and July, then increasing again for releases in August.

Spring and summer upwelling has been shown to influence marine survival of Oregon coho stocks (Nickelson 1986). Better upwelling generally means better marine survival on an annual basis, but the mechanisms relating marine survival to the timing of seawater entry are unknown. The coho adult and spring chinook jack returns from 1985 releases show the same pattern; excellent survival from

Table 4. *Adult return of spring chinook releases at Coos Bay, Oregon*

Year of release	Date of release	Size of release (g)	Return (%)
1979	11/06	45.3	.357
	10/31	39.7	.166
	10/30	38.2	.036
1981	07/15	23.1	.964
	07/15	16.3	.540
	08/14	37.0	1.012
	08/18	35.7	.572
	09/01	37.8	.988
	09/01	30.0	.790
	10/09	35.5	.099
	10/09	30.5	.179
1982[a]	09/04	47.0	.888
	09/04	46.0	.787
	08/27	46.0	.846
	08/27	40.0	.654

[a] Through 4 year old returns only.

May and August releases, poor survival from June and July releases. Ocean conditions not only determine interannual survival but intra-annual survival as well.

SIZE AT RELEASE

In 1985 two groups of coho were released each month from May through August. In each month the groups were released within 24 hours of each other. They were from the same stock and had the same rearing history. The only difference between the groups was that they varied in size at release (Table 3), a difference which varied from 6 to 13 grams. In all four months the larger fish returned at a higher rate. The difference in rate of return varied from 26 to 231 percent. The same pattern was observed for 1981, 1982, and 1984 releases.

In the period 1979-82 several coded-wire tagged groups of spring chinook were released over several months each year (Table 4). In five of six comparisons the larger fish survived at a higher rate than the smaller fish. It appears that for a given time of release a large fish has a better chance of survival than a small fish.

MANIPULATIONS AT THE TIME OF RELEASE

Three release methods have been evaluated in the last 3 years: forced versus self, ladder versus tube, and night versus day.

The forced versus self-release was evaluated as follows. There were four replicates or release periods (Table 5) each with a self-released and a forced release group. Self-release consisted of removing the screen at the end of the raceway and allowing the fish to outmigrate over a 2-week period. At the end of that period any fish remaining in the raceway were crowded out. In the middle of the 2-week period another raceway of fish with an identical rearing history was crowded out at one time. The procedure was repeated four times over 4 months. In all four replicates the forced release fish survived better than the self-released group. This result is consistent with the work of Wagner (1968) on the Alsea and of Ewing (Oregon Department of Fish and Wildlife, personal communication) on the Rogue rivers, both of whom worked with steelhead.

Another release comparison was night versus day release. In that trial one group of two tag codes was released during the day; a second group of two tag codes was released that night (Table 6). The following day the release groups were repeated. The results were surprising. For both replicates, fish released in daylight survived better than fish released at night. Fish released at night were much more docile: they did not resist being crowded down the ladder and appeared to leave the site in much better condition. Fish released during the day typically fight the crowding screen to exhaustion, finally going down the ladder unable to maintain an upright position; yet the day release outperformed the night release. I can offer no explanation as to why this should be.

Three groups of fish of two tag codes each were involved in a ladder versus tube release experiment. One group was released down an open ladder, one group was released through an enclosed tube, and one group was transferred by tube into a net pen, held for 48 hours, and then released (Table 7). All treatment groups were released into the estuary within one hour of each other. The

Table 6. *Adult percentage return of coho salmon released in daylight and at night, Coos Bay, Oregon in 1984 and returning in 1985.*

	Date of release	
	June 18	June 19
Daytime release	.96	1.62
	1.00	1.88
Mean	.98	1.75
Nighttime release	.84	.96
	.80	1.06
Mean	.82	1.01

Table 7. *Adult percentage return of coho released by three different methods at Coos Bay, Oregon in 1984 and returning in 1985.*

	Method A	Method B	Method C
Percentage return	.77	.58	.73
	.61	.46	.61
Mean	.74	.52	.72

Notes: Method A: Released down ladder.
Method B: Reared in raceway and released down tube.
Method C: Reared in raceway, moved by tube to net pen, held 48 hours, and released.

results were not significantly different, but the ladder and tube/pen releases did survive at a slightly higher rate than the tube release.

In two of the three release methods investigated, then, significant effects were found to result from the different manipulations. Forced release fish survived better than self-released fish. Fish released in daylight survived better than fish released at night. In both cases, adult survival varied by a factor of two.

BARGING

If precocious males (jacks) can be used to predict adult survival then it has been postulated that the major variability in marine survival must occur prior to the time some fish return as jacks, i.e., the first 3-6 months of ocean residence. One theory as to why some release periods are better than others relates to the abundance and feeding habits of nearshore predators during smolt outmigration. Placing fish in a transport device and towing them a short distance offshore prior to releasing them into the ocean might prevent them from being

Table 8. *Comparison of return percentage of tagged coho released onshore and offshore at Coos Bay, Oregon, in 1985 and returning in 1986.*

Month of release	Percentage of fish released		
	Onshore	7 miles from shore	14 miles from shore
May	5.95 (2)[a]	4.36 (4)	4.45 (4)
June	3.79 (2)	3.21 (4)	2.71 (4)
July	2.53 (2)	.74 (8)[b]	
August	10.27 (2)	9.21 (4)	8.44 (4)

[a] ( ) = Number of tag codes.
[b] Barge door opened prematurely at 6 miles in rough sea conditions.

subjected to that nearshore predation. The barge trials are, therefore, a test of the predator hypothesis. The experimental design for the 1985 Anadromous offshore releases is shown in Table 8, along with return rates. Initial results are not encouraging. It is important to remember that two questions are being addressed. The first is the question whether an offshore release will improve survival. The second is the success of the transport device itself.

In earlier studies, Oregon Aqua Foods towed smolts offshore from Yaquina Bay in a net pen. They consistently obtained 2:1 survival ratios of offshore over estuary release. After 2 years of experience with the barge itself we question the suitability of the basic design. In anything other than a calm ocean fish were blown out the top of the barge by the surge of swells passing through it. On some trips up to 10 percent of the fish were found dead in the barge when it returned from an offshore release. Other fish were trapped by the internal doors used to separate tagged groups of fish. In 1987 we used a modified transport device in an attempt to reduce stress on fish while the device is under tow.

DISCUSSION AND CONCLUSIONS

Cultural practices have not significantly influenced marine survival of either coho or spring chinook released by Anadromous. That does not mean that cultural techniques are unimportant but rather that they need to be applied with the objective of maximizing production in an economic fashion, not with the expectation of affecting marine survival. Fish have to be kept healthy and growing; a fish which is sick and small when it is released will obviously not return as an adult.

Techniques designed to meet conditions external to the facility can influence estuarine and marine survival. In 1985 fish released by Anadromous in May and August survived at a rate of four times greater than fish of similar size and rearing histories released in June and July. Forcing fish out of a facility rather than letting them self-migrate increased survival in all four replicates. Day releases produced greater survival than similar fish released by the same method at night, but this should not be considered an axiom. For a given time of release a larger fish has a higher probability of marine survival. The fish released by Anadromous are

much larger than needed for seawater adaptation. Coho, for example, are two to four times larger than needed for successful seawater introduction.

Why do these manipulations work? The simple answer is that we don't know. The marine environment determines ultimate survival from smolt to adult; yet it has not been investigated much beyond attempts at finding simple associations among a limited set of variables. Mechanisms controlling marine survival involve complex interactions among many and mostly undefined processes. We need to attempt to understand those processes which explain the role of hatchery practices in modifying marine survival. An understanding of those processes can provide the basis for hatchery practices designed to increase marine survival.

LITERATURE CITED

Bilton, J.T. 1978. Returns of adult coho salmon in relation to mean size and time of release of juveniles. Can. Tech. Rep. of Fish. and Aquat. Sci. 832:73 p.

Buckman, M., and R.D. Ewing. 1982. Relationship between size and time of entry into the sea and gill (Na+K)—ATPase activity for juvenile spring chinook salmon. *Trans. Amer. Fish. Soc.* 111:681-687.

Evenson, M.D. and R.D. Ewing. 1984. Cole Rivers hatchery evaluation. Oregon Dept. Fish Wildl. Ann. Prog. Rep. 58 p.

Gowan, R.E. and W. McNeil. 1983. Factors associated with mortality of coho salmon (*Oncorhynchus kisutch*) from salt water release facilities in Oregon. In W.G. Pearcy, ed., *The Influence of Ocean Conditions on the Production of Salmonids in the North Pacific.* Oregon State University Sea Grant College Program, Corvallis. (ORESU -W-83-001.)

Hager, R.C. and R.E. Noble. 1976. Relation of size at release study. Columbia River study analysis and documentation completion report. Salmon Culture Division. Washington Dep. of Fish., Olympia.

Hansen, L.P. 1982. Salmon ranching in Norway. In *Sea ranching of Atlantic salmon.* Cost 46/4 workshop. Lisbon, Portugal.

Johnson, S.L. 1982. A review and evaluation of release strategies for hatchery reared coho salmon. 82-5. Oregon Dept. Fish Wildl.

Nickelson, T. 1986. Influences of upwelling, ocean temperature, and smolt abundance on marine survival of coho salmon (*Oncorhynchus kisutch*) in the Oregon Production area. *Can. J. Fish. Aquatic Sci.* 43:527-535.

Perry, T. 1983. Bird predation of hatchery released chinook salmon. Proc. N.W. Fish Cult. Conf. Moscow, ID. pp 76-77.

Wagner, H. 1968. Effect of stocking time on survival of steelhead trout *Salmo gairdnerii* in Oregon. *Trans. Am. Fish. Soc.* 98:374-379.

Chum Salmon as Indicators of Ocean Carrying Capacity

Ernest O. Salo

Professor Emeritus
Fisheries Research Institute
University of Washington

Chum salmon typically exhibit cyclical variations in life history characteristics such as survival, abundance, and size and age at maturity. Puget Sound chum salmon, for example, exhibit strong alternate year variations in abundance (Figures 1 and 2). An example of regular variations in marine survival is provided by Fraser River chum salmon (Figure 3). Variations in size at maturity are illustrated in Figure 4, using data from two fisheries in Puget Sound. Variations in age at maturity are shown for Puget Sound and the Fraser River (Figures 5 and 6, respectively). More detailed information on variations in life history patterns of chum salmon is contained in Salo (in press).

INTERACTIONS WITH PINK SALMON

Soviet (Smirnov 1947, Lovetskaya 1948) and North American (Rounsefell and Kelez 1938, Noble 1955) scientists have noted the alternation in abundance of chum salmon between odd and even years and speculated about the cause. Recent research indicates density-dependent effects in the marine environment for some salmon species (Rogers 1980, Beacham and Starr 1982, Peterman 1984). Such effects were believed to occur early in the marine

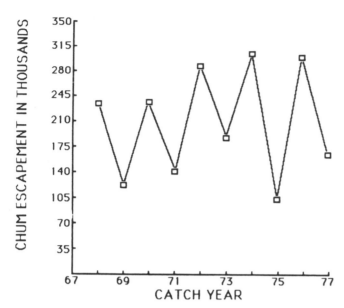

Figure 1. Escapements of chum salmon to Puget Sound, 1968-77.
After Gallagher 1979.

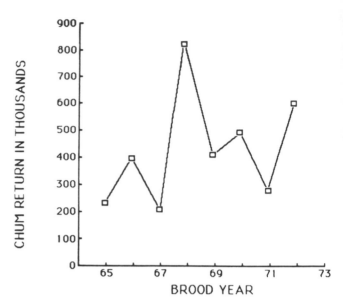

Figure 2. Returns of chum salmon to Puget Sound by brood year.
From Gallagher 1979.

environment for both chum and pink salmon (Birman 1960, Gallagher 1979). Beacham and Starr (1982) suggest that marine survival of chum salmon is influenced by the abundance of adjacent year classes. There is a good possibility that chum and pink salmon interactions in the marine environment have important effects on life history characteristics of chum salmon. Figure 7 compares abundance of North American pink and chum salmon stocks.

In years when pink salmon juveniles are abundant, the feeding and growth rates of the chum salmon have been reported to be lessened and their diets changed (Ivankov and Andreyev 1971). In Puget Sound stocks, the age of maturation of chum salmon shifts to a pattern that increases reproductive potential in alternate years, coinciding with years of scarcity of (odd-year) pink salmon (see Figure 1). These responses may have evolved to minimize competition between pink and chum salmon (Gallagher 1979, Smoker 1984). Gallagher (1979) concluded that the shift of reproductive effort of chum salmon to coincide with years when pink salmon are scarce represents a genetic adaptation to minimize competition. Smoker (1984) developed a model of chum-pink salmon interactions based on two stocks. He concluded that a fairly strong genetic mechanism must be involved.

Consistent fluctuations in age at maturity of chum salmon also occur in some stocks that coexist

with both even-year and odd-year pink salmon. Chum salmon in Olsen Creek, Alaska, which has both even- and odd-year pink salmon stocks, show a more complicated pattern than do those in Puget Sound where there are only odd-year pink stocks (Gallagher 1979). A complicated but fairly regular pattern exists in age at maturity for Bolshaya River chum salmon in western Kamchatka, U.S.S.R., where the dominant pink salmon cycles periodically alternate between odd and even years. The presence of any regular pattern in Bolshaya chum

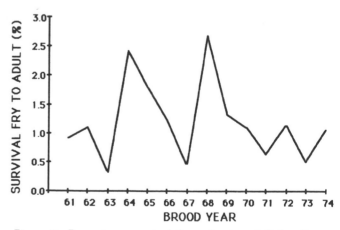

Figure 3. Percentage survival from fry to adult for Fraser River chum salmon for brood years 1961-74.
Data from Beacham and Starr 1982.

salmon cannot be easily explained by a genetic model, possibly because of these periodic changes in the dominant cycles of pink salmon (Semko 1954, Sano 1966). The even-year broods produce fewer chum salmon in the Bolshaya River than the odd-year broods, but there is not the expected shift of age distribution to the odd-brood years (Salo, in press).

With substantial increases in the numbers of juvenile chum salmon released from the hatcheries

on Hokkaido Island, Japan, one might expect increased competition with pink salmon. Data on age composition of chum adults returning to Hokkaido have not been analyzed along with possible year-class dominance of any interacting pink salmon; however, catches of chum salmon in odd years have been consistently larger than those in even years.

OCEAN CARRYING CAPACITY

The total catch of north Pacific chum salmon from 1925 through 1981 averaged over 41 million fish (range 18 to 83 million). In this period the mean Asian catch was 26.5 million while the mean North American catch was 11.5 million (Figure 8). In the period 1952-82 the mean annual harvest of chum salmon included 24.9 million from Japan, 6.9 million each from the U.S.S.R. and the U.S., and 2.4 million from Canada.

The decade with the greatest total catch of chum salmon was 1934-43, with an annual average harvest of 57.5 million fish. If the fish had an average weight of 3.6 kg, as the catch data suggest, the average catch would have weighed 210,600 mt. The peak year was 1936, when 83 million fish weighing 304,000 mt were harvested. An assumed

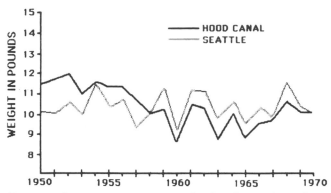

Figure 4. Average purse seine weights for chum salmon from selected areas in Puget Sound, 1950-70.
From Pratt 1974.

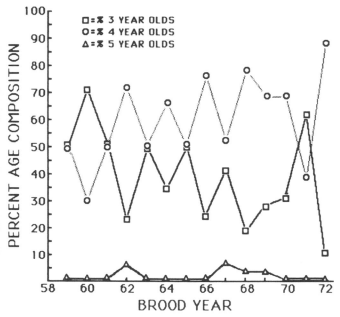

Figure 5. Age at return of Puget Sound chum salmon by brood year 1959-72.
From Gallagher 1979.

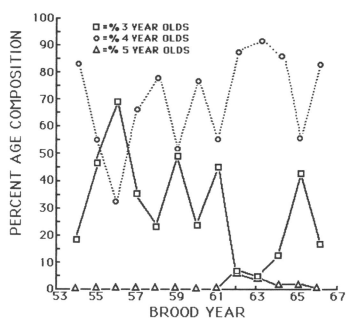

Figure 6. Age at return of chum salmon to Fraser River, B.C. by brood year 1954-66.
From Gallagher 1979.

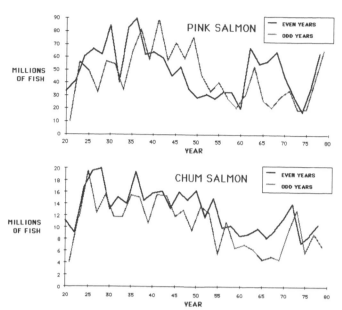

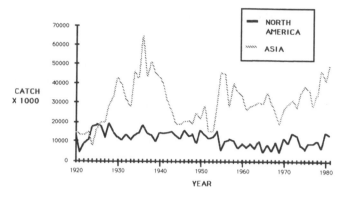

Figure 8. *Total catch of chum salmon of Asian and North American origins by commercial fisheries of the U.S.S.R., Japan, U.S., and Canada, 1920-77.*
From Salo (in press).

Figure 7. *North American catches of chum and pink salmon, 1920-80.*
After Harris 1980.

catch escapement ratio of 2:1 would imply an annual production of 126 million mature fish (462,000 mt). Using the estimated age composition for the catch (Salo, in press), this would mean that the total biomass of immature and mature chum salmon (standing stock) in the north Pacific would have weighed between 860,000 and 1,300,000 mt. Assuming marine survival to be about 3 percent, approximately four billion chum fry were recruited annually into the north Pacific during peak years of natural production (1936-38).

Variations in life history patterns of chum salmon suggest that the species responds to competition by altering certain vital characteristics including abundance and size and age at maturity. The implication is clear: the capacity of the north Pacific to grow salmon is not unlimited. The six species of Pacific salmon (including masu salmon) may adapt to competition by, as it were, partitioning the ocean in space and time. Interactions between pink and chum salmon suggest that chum salmon, because of considerable variability in life history patterns, are able to adapt to competition from alternate-year cycles of scarcity and abundance of pink salmon.

Large-scale hatchery production of chum salmon in Japan, where annual release numbers now exceed two billion juveniles, should be carefully monitored

for possible impacts on ocean survival, and age and size at maturity. Historic high levels of annual recruitment of chum juveniles to the north Pacific were calculated earlier in this report to be about four billion. This level is once again being approached with increasing numbers of juveniles being released from hatcheries. However, the distribution of production has shifted geographically, and ocean distribution patterns of feeding fish may be changing. It will be interesting to see if the capacity of the ocean to grow chum salmon and possibly other species will soon be reached.

LITERATURE CITED

Beacham, T.D., and P. Starr. 1982. Population biology of chum salmon, *Oncorhynchus keta*, from the Fraser River, British Columbia. *Fishery Bulletin* 80(4): 813-825.

Birman, I.B. 1960. New information on the marine period of life and the marine fishery of Pacific salmon. Ikhtiologichesbaia kimissiia akadencii nauk SSSR, trudy saveshchanii, No. 10, pp 151-164. Moscow.

Gallagher, Jr., A.F. 1979. An analysis of factors affecting brood year returns in the wild stocks of Puget Sound chum (*Oncorhynchus keta*) and pink salmon (*Oncorhynchus gorbuscha*). M.S. Thesis, Univ. Washington, Seattle. 152 p.

Ivankov, V.N. and V.L. Andreyev. 1971. The South Kuril chum [*Oncorhynchus keta* (Walb.)]—Ecology, population structure and the modeling of the population. *J. Ichthyol.* 11(4):511-524.

Lovetskaya, E.A. 1948. Data on the biology of the Amur chum salmon. Izv. Tikhookean. Nauchno-issled. Inst. Rybn. Khoz. Okeanogr. 27:115-137. In: *Pacific Salmon: Selected articles from Soviet periodicals.* Transl. from Russian, IPST Cat. No. 341 (1961):101-126.

Noble, R.E. 1955. Minter Creek Biological Station progress report Fall 1955. State of Washington Dept. of Fisheries (mimeo).

Peterman, R.M. 1984. Interaction among sockeye salmon in the Gulf of Alaska. Pages 187-199 in *The Influence of Ocean Conditions on the Production of Salmonids in the North Pacific.* Oregon State University Sea Grant College Program, Corvallis. (ORESU-W-83-001.)

Rogers, D.E. 1980. Density-dependent growth of Bristol Bay sockeye salmon. Pages 267-283 in W.J. McNeil and D.C. Himsworth, eds., *Salmonid Ecosystems of the North Pacific.* Oregon State University Press, Corvallis.

Rounsefell, G.A., and G.B. Kelez. 1938. The salmon and salmon fisheries of Swiftsure Bank, Puget Sound, and the Fraser River. *Bull. U.S. Bur. Fish.* 48:693-823.

Salo, E.O. (In press). The life history of chum salmon (*Oncorhynchus keta*). Scientific Information and Publications Branch, Canada Dept. of Fisheries and Oceans.

Sano, S. 1966. Chum salmon in the Far East. *Int. North Pac. Fish. Comm., Bull.* 18:41-58.

Semko, R.S. 1954. The stocks of West Kamchatka salmon and their commercial utilization. Izv. Tikhookean. Nauchno-issled. Inst. Rybn. Khoz. Okeanogr. Vol. 41, p. 3-109, 20 figs., 65 tables. 1960. (Transl. from Russian by Fish. Res. Board Canada, Biol. Sta., Nanaimo, Brit. Columbia, Transl. Ser. 288).

Smirnov, A.I. 1947. Condition of stocks of the Amur salmon and causes of the fluctuations in their abundance. Izv. Tikhookean. Nauchno-issled. Inst. Rybn. Khoz. Okeanogr. 25:33-51. In *Pacific Salmon: Selected articles from Soviet periodicals.* Transl. from Russian. IPST Cat. No. 341(1961):66-85.

Smoker, W.W. 1984. Genetic effect on the dynamics of a model of pink and chum salmon. *Can. J. Fish. Aquat. Sci.* 41:1446-1453.

Bristol Bay Smolt Migrations
Timing and Size Composition and the Effects on Distribution and Survival at Sea

Donald E. Rogers
Research Professor
Fisheries Research Institute
University of Washington

INTRODUCTION

Bristol Bay (see Figure 1) produces the largest number of Pacific salmon in North America. The average annual run of adult salmon during the most recent 10-year period (1977-86) numbered about 44 million fish, which was about 20-25 percent of the average run to all North American rivers combined (Rogers 1987). Sockeye salmon constituted 76 percent of the recent Bristol Bay salmon runs and 93 percent of the sockeye were bound for rivers in inner Bristol Bay from Ugashik to Togiak (Table 1).

Since 1978, salmon abundance in Bristol Bay has probably been as high as or higher than at any other time since the inception of commercial fisheries nearly 100 years ago. The causes of interannual variation in the abundance of adult salmon remain one of the most important questions for salmon management and it has been suggested that the recent increase in salmon abundance was at least partly caused by an increase in marine survival associated with warmer temperatures (Rogers 1984).

Marine survival is believed to be lowest for salmon during their first few months at sea when they are in estuaries or near shore and their body size is relatively small, less than 15 cm long (Ricker

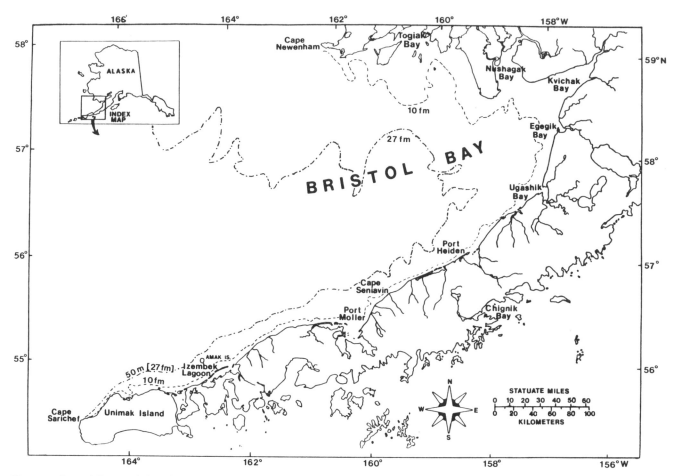

Figure 1. Bristol Bay and the Alaska Peninsula.

1976, Hartt 1980). Sockeye stocks with relatively large smolts tend to have higher marine survivals than those with small smolts; however, for individual stocks there may be little correlation between interannual variation in body size and marine survival (Mathews 1984). Recently there has also been concern that marine survival may be density-dependent, with survival reduced at high densities of seaward migrants. This would presumably be caused by reduced growth and increased mortality from starvation or size-dependent predation.

The purpose of this paper is to describe some aspects of the Bristol Bay smolt migrations (abundance, body size, timing, and temperature) with emphasis on interannual variation in the distribution of smolts in Bristol Bay and their marine survival based on subsequent adult returns. Although the factors that may cause variation in the survival

of Bristol Bay stocks may be applicable to other salmonid stocks, the biological and physical environment in Bristol Bay is unique. Thus the effects of these factors may be expressed quite differently in other areas. However, Bristol Bay does provide one case history for the major salmonid stocks of the north Pacific.

THE SMOLT MIGRATION

The annual outmigration of sockeye salmon smolts from four Bristol Bay watersheds (Wood, Kvichak, Naknek and Ugashik, Figure 2) was monitored by fyke-net catches during the 1950s and 1960s. The purpose was to estimate the relative abundance of smolts to forecast subsequent adult runs. These projects were partially successful but they generally did not provide direct estimates of

Table 1. *Catches (C) and escapements (E) of salmon in Bristol Bay and the north side of the Alaska Peninsula.* (In millions of fish)

		Year of Run 1976	1977	1978	1979	1980	1981	1982	1983	1984	1985	1986
Sockeye												
East side[a]	C	4.0	4.0	6.3	17.6	17.2	17.4	8.6	31.4	22.2	21.9	12.8
	E	4.2	3.4	6.2	15.2	29.8	5.7	4.7	6.4	14.4	11.3	5.5
West side[b]	C	1.6	0.8	3.6	3.8	5.0	8.3	6.6	5.9	2.5	1.5	3.0
	E	1.7	1.4	3.8	3.3	8.8	3.1	2.3	2.1	2.0	1.9	2.4
North Peninsula	C	0.6	0.5	0.9	2.0	1.4	1.8	1.4	2.1	1.7	2.6	2.4
	E	0.5	0.5	1.2	1.6	1.4	1.3	0.7	0.6	0.8	0.9	0.6
Total run		12.6	10.6	22.0	43.5	63.6	37.6	24.3	48.5	43.6	40.1	26.7
Chum												
East side[a]	C	0.4	0.4	0.2	0.2	0.3	0.5	0.3	0.6	0.8	0.4	0.4
	E	—	—		—	—	—	—	—	—	—	—
West side[b]	C	1.0	1.2	0.9	0.7	1.1	1.0	0.6	0.9	1.0	0.5	0.7
	E	0.9	1.1	0.7	0.5	1.4	0.5	0.3	0.3	0.6	0.5	—
North Peninsula	C	0.1	0.1	0.2	0.1	0.7	0.7	0.3	0.3	0.8	0.7	0.3
	E	0.3	0.7	0.3	0.3	0.8	0.5	0.5	0.4	0.9	0.3	0.2
Total run[c]		3.1	3.8	2.5	2.0	4.8	3.4	2.2	2.6	4.6	2.6	2.4
Pink												
East side[a]	C	0.3	*	0.7	*	0.3	*	0.1	*	0.2	*	0.1
	E	—	—	2.0	—	0.4	—	0.2	—	1.3	—	0.4
West side[b]	C	0.8	*	4.4	*	2.3	*	1.4	*	3.2	*	0.3
	E	0.9	—	9.5	—	2.9	—	1.7	—	3.2	—	0.1
North Peninsula	C	*	*	0.5	*	0.3	*	*	*	*	*	*
	E	*	*	0.1	*	0.1	*	*	*	0.1	*	—
Total run		2.4	*	17.2	*	6.3	*	3.5	*	8.0	*	0.8
Coho												
Total run[d]		0.1	0.2	0.4	0.6	0.8	1.0	1.4	0.4	1.3	0.8	0.7
Chinook												
Total run[e]		0.2	0.2	0.4	0.4	0.3	0.5	0.5	0.5	0.3	0.3	0.1

Sources: Various ADF&G reports e.g., Yuen et al. (1986) and Schaul et al. (1985) and D. Eggers (ADF&G) for 1986 statistics.

[a] East side is Naknek-Kvichak, Egegik, and Ugashik Districts.

[b] West side is Nushagak and Togiak Districts.

[c] Includes estimates of escapements for east side from rate of exploitation on sockeye.

[d] Runs estimated by 2 x catch (mostly west side).

[e] West side only; other runs are negligible.

* Less than 100,000 fish.

marine or post-smolt survival; however, they did provide important information on timing, age, and size composition in the migrations (Eggers and Rogers 1978). In 1971, the Alaska Department of Fish and Game (ADF&G) initiated sonar-based enumeration coupled with fyke-net sampling to annually estimate the number, age, and size of the seaward migrants from the Kvichak system. This method was employed in Wood River beginning in 1975 and then in most of the other Bristol Bay systems in 1982. Annual statistics on the sockeye migrations are presented in various ADF&G reports (e.g., Eggers and Yuen 1984).

Information on the outmigrations of salmonids other than sockeye from Bristol Bay rivers is either totally lacking or available for a limited number of years; for example, Nushagak pink salmon migrations in 1981 and 1983 (Mesiar 1986). Based on the adult runs, the majority of seaward migrants other than sockeye migrate from the Nushagak river system (see Figure 2), which is the largest river system flowing into Bristol Bay.

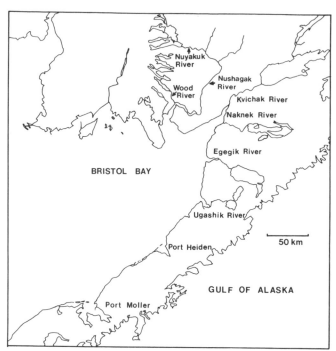

Figure 2. Bristol Bay rivers with smolt enumeration projects.

Sockeye salmon smolts begin their migrations in mid-May to early June and the migrations are of short duration from single lakes (e.g., 2-3 weeks from Egegik) but continue longer in systems comprising several lakes (e.g., about 2 months from Wood River). The majority of smolts leave the lakes during the brief period of darkness or semi-darkness (about 4 hours a day at this time of year). The migrations are characterized by extreme fluctuations in numbers from night to night, especially from single lakes, where the majority of smolts in the season's migration may leave on just a few nights. Smolts that have spent 2 years in the lake and average about 10-12 cm in length tend to migrate earlier in the spring than smolts that have spent 1 year in the lake and average about 8-10 cm in length. The average body size of smolts for a given age changes little during the course of the migration except at Wood River. The late migrants from this system typically increase in size after mid-July as a result of spring growth. The very late migrants are often larger than the earliest by about 1 cm.

Direct information on salmonid migrations through Bristol Bay is largely limited to (1) their general distribution, which is a migratory route mainly within 40 km of the Alaska Peninsula shore-line; (2) the approximate swimming speed of sockeye (about one body length per second); and (3) their occurrence in Bristol Bay from mid-May through at least September (Straty 1974, Bax 1985, Isakson et al. 1986). Although we lack interannual measurements of the migrations of the various stocks through Bristol Bay, we can approximate these migrations from the numbers, size, and timing of the migrations from the lakes and the distances to various points along the migratory route.

There is little available information on interannual variation in the physical and biological conditions in Bristol Bay during the period of salmon migrations other than temperatures. Nushagak and Kvichak bays are characterized by large tidal range (7-8 m), strong currents, very turbid water, extensive mud flats, and low salinities. The net flow of water from these bays is to the northwest, south of Cape Newenham. There is little available food for salmon and visibility is low; hence, there is little growth in the inner bays (Straty and Jaenicke 1980). One thousand to fifteen hundred belukha whales congregate in Nushagak and Kvichak bays with the departure of winter ice in April. They feed on salmon smolts (utilizing their sonar capabilty) until the arrival of adult sockeye salmon in late June. Their feeding then turns to the adults (Frost et al. 1983).

No other significant predation is known to occur in the inner bay and the whales are apparently absent from Egegik and Ugashik bays. Seaward of Ugashik Bay, the coastal waters become more oceanic. There is a coastal current with a flow into Bristol Bay of 1-2 cm/sec (Schumacher and Moen 1983). Food is more plentiful in the outer bay and salmon begin to grow there (Straty and Jaenicke 1980). Fish surveys by Isakson et al. (1986) in 1984-85 caught fifty species. Most were juveniles or small fish less than 20 cm long; Pacific sand lance were especially abundant, often more numerous than juvenile salmon in the epipelagic and nearshore waters. Juvenile pink and chum salmon were concentrated nearshore, whereas the larger juvenile sockeye and coho salmon were more abundant offshore. Large fishes other than salmon were relatively scarce in the coastal surface waters. The main migratory route of adult salmon is offshore about 25 to 150 km into the vicinity of Ugashik Bay (Straty 1975). The coastal waters along the north side of the Alaska Peninsula contain rather large populations of fish-eating sea birds; however, no significant predation on juvenile salmon has been identified from stomach content analysis (J. Sanger, U.S. Fish and Wildlife Service, personal communication).

During the course of their seaward migrations in Bristol Bay, salmon have encountered a wide range of temperatures over the years, ranging from 2°C to 14°C. Year-to-year variation in temperature affects the timing of the smolt migrations from the rivers and, in addition, can affect their rate of travel and growth (Brett and Glass 1973). Temperature may also affect the vulnerability of salmon to predators.

Very little is known about the seaward migration beyond Port Moller. The following year, in July and August, immature Bristol Bay sockeye are predominant in the coastal waters south of the Aleutians where they have been sampled south of Adak annually since 1956 (Harris and Rogers 1979, Takagi and Ito 1986). Maturing Bristol Bay sockeye are distributed over a broad area from Central Gulf of Alaska to about 165°E (French et al. 1976), and yet the majority return to inner Bristol Bay within a 2-week period.

ABUNDANCE OF SMOLTS

Until recently, the annual abundance of Bristol Bay smolts could only be estimated from the adult returns by year of seaward migration and average marine survival based on a few observations (Rogers 1978). The sockeye salmon returns to Bristol Bay from the 1954 to 1973 migrations were very cyclical and this was mostly caused by the returns to the largest lake system (Kvichak); however, the returns to the other eastside systems (Naknek, Egegik, Ugashik) were also relatively large in the peak cycle years of the Kvichak system (Figure 3). I estimated that the annual numbers of sockeye salmon migrating into inner Bristol Bay ranged from about fifty million to six hundred and fifty million during this period. The cycle changed after 1975 with two returns rather than one large return to the Kvichak system and significantly larger returns to the other lake systems.

The numbers of sockeye salmon smolts in the migrations since 1975 were estimated by adding the numbers enumerated in rivers with sonar projects to estimates of numbers in rivers without sonar projects. Estimates for eastside rivers with-

out sonar projects were made by dividing the adult returns to these by the marine survival for the Kvichak River. Estimates for other Nushagak-Togiak rivers used marine survival data for Wood River (Table 2). The estimated numbers of smolts for other eastside rivers may have been somewhat too high and those for other westside rivers somewhat too low if marine survival was size-dependent, since Kvichak smolts were generally smaller than those from other eastside systems (Table 3).

The estimated annual numbers of sockeye salmon smolts in migrations since 1975 ranged from about 220 million to 640 million, and marine survival ranged from a low of 3 percent (1980) to a high of 13 percent (1978). The adult returns generally corresponded to the numbers of smolts except for the low returns from the 1980 migration, especially from the Kvichak system (Figure 4). Assuming that the other species of salmon had marine survivals comparable to Wood River sockeye stocks, there were an additional 50 to 60 million seaward migrants in even-numbered years and 130 million (1983) to 380 million (1977) in odd-numbered years.

Mesiar (1986) estimated that marine survivals for pink salmon in the 1981 and 1983 migrations were 3.5 percent and 9.9 percent, respectively. Survivals for Wood River sockeye in those years were

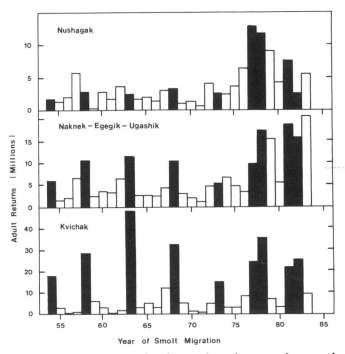

Figure 3. Adult returns of sockeye salmon by year of seaward migration, 1954-83.

Table 2A. Sockeye salmon smolts migrating from and adults returning to Kvichak River.

	Smolts			Adult returns			
Year of migration	Number (millions)	Percent age 1.	Mean wt. (g)	Number (millions)	Percent age 1.	Percent age .3	Survival (%)
1971	92	93	6.1	.59	68	69	0.6
1972	55	2	9.9	4.72	1	11	8.6
1973	197	3	8.2	14.98	2	4	7.6
1974	27	11	13.2	2.64	11	10	9.8
1975	—	37	11.4	2.89	52	48	—
1976	111	97	6.0	8.33	91	29	7.5
1977	193	40	8.2	24.07	28	8	12.5
1978	246	13	7.6	34.83	17	4	14.1
1979	55	53	8.0	6.54	38	12	11.9
1980	193	94	6.1	2.90	91	39	1.5
1981	252	87	6.0	21.81	91	13	8.7
1982	240	62	6.6	24.66	17	19	10.3
1983	83	8	8.2	9.17	11	7	11.0
1984	89	58	8.2	1.35+	—	—	—
1985	24	92	5.6	—	—	—	—

Table 2B. Sockeye salmon smolts migrating from and adults returning to Wood River.

	Smolts			Adult returns			
Year of migration	Number (millions)	Percent age 1.	Mean wt. (g)	Number (millions)	Percent age 1.	Percent age .3	Survival (%)
1975	34	83	6.0	1.29	95	83	3.8
1976	106	95	4.5	4.52	98	35	4.3
1977	74	83	4.5	3.91	88	49	5.3
1978	55	85	4.3	5.39	81	53	9.8
1979	66	92	7.8	3.81	80	62	5.8
1980	48	96	4.1	2.37	97	48	4.9
1981	97	66	7.0	5.04	87	33	5.2
1982	37	87	5.2	1.46	97	66	3.9
1983	24	83	6.9	1.94	90	64	8.1
1984	24	97	7.8	.58+	—	—	—
1985	36	88	7.1	—	—	—	—

Table 2C. Sockeye salmon smolts migrating from and adults returning to Naknek River.

	Smolts			Adult returns			
Year of migration	Number (millions)	Percent age 1.	Mean wt. (g)	Number (millions)	Percent age 1.	Percent age .3	Survival (%)
1982	128	90	8.9	3.19	64	55	2.5
1983	53	70	9.3	6.02	63	63	11.4
1984	81	40	10.5	.64+	—	—	—
1985	20	32	10.8	—	—	—	—

Table 2D. Sockeye salmon smolts migrating from and adults returning to Egegik River.

	Smolts			Adult returns			
Year of migration	Number (millions)	Percent age 1.	Mean wt. (g)	Number (millions)	Percent age 1.	Percent age .3	Survival (%)
1982	63	78	10.6	7.35	38	49	11.7
1983	19	11	13.2	6.71	22	27	35.3
1984	49	35	11.5	4.35+	—	—	—
1985	66	83	11.5	—	—	—	—

Table 2E. Sockeye salmon smolts migrating from and adults returning to Ugashik River.

	Smolts			Adult returns			
Year of migration	Number (millions)	Percent age 1.	Mean wt. (g)	Number (millions)	Percent age 1.	Percent age .3	Survival (%)
1982	—	—	8.2	5.15	63	51	—
1983	44	70	9.3	7.99	50	41	18.2
1984	158	48	8.6	2.61+	—	—	—
1985	34	37	10.5	—	—	—	—

Table 2F. Sockeye salmon smolts migrating from and adults returning to Nuyakuk River.

	Smolts			Adult returns			
Year of migration	Number (millions)	Percent age 1.	Mean wt. (g)	Number (millions)	Percent age 1.	Percent age .3	Survival (%)
1982	—	—	—	.59	98	78	—
1983	30	96	4.8	2.15	91	85	7.2
1984	7	86	5.2	.10+	—	—	—
1985	23	97	5.6	—	—	—	—

not significantly different (5.2 percent and 8.1 percent). Based on the Wood River sockeye survival for 1977 (5.3 percent) and the return of pink salmon from that year of migration (16.6 million, excluding North Peninsula stocks), we can estimate that about 310 million pink salmon migrated to sea in 1977. Combined with 540 million sockeye and about 70 million smolts of other species, the 1977 seaward migration of about 920 million was probably the largest in recent years. However, these fish do not all migrate to sea from the same place nor at the same time, and during the course of their migration out of Bristol Bay the various stocks and species may be at least partially separated. Large concentrations during seaward migration may be an advantage for survival from predation, but a disadvantage for growth (if food is scarce) and perhaps for survival at a later time.

TIMING OF MIGRATIONS

Spring warming begins earlier on the east side than the west side of Bristol Bay because the Alaska Peninsula is more influenced by a maritime climate (see Figure 2); thus, smolt migrations begin earlier from lakes on the east side. To examine year-to-year variation in the timing of smolt migrations, four measurements were taken of spring weather during 1955-86: air temperature (April-June), water temperature in the Kvichak River (May 25-June 7) and Wood River (June 1-15), and date of ice breakup in Lake Aleknagik (Wood River Lakes). These measurements were variously correlated, and thus the four relative deviations (annual observations minus the mean, divided by the standard deviation) were averaged to obtain an annual temperature index (Figure 5).

There was no continuous record of water temperature in Bristol Bay; however, annual nearshore temperatures at Kodiak in the Gulf of Alaska were available except for 1965-66; estimates for the latter year were made from offshore temperatures (Table 4). The interannual variation in Kodiak April-June temperatures generally corresponded to the Bristol Bay indices with two notable exceptions. Temperatures were very warm in Bristol Bay lakes but colder than average at Kodiak in 1974 and very warm at Kodiak but just below average in Bristol Bay lakes in 1977. Based on the limited temperature data for Bristol Bay (Port Heiden to Port Moller), smolts migrating into Bristol Bay usually encountered similar or cooler temperatures (Table 5). An exception was in 1985 when smolts encountered much warmer temperatures in the bay. Water temperatures were probably above average along the coast of the Alaska Peninsula in 1985, whereas the temperature index was one of the coldest.

The timing of annual sockeye smolt migrations was estimated by the dates by which 10 percent, 50 percent, and 90 percent of the season's migration had left the lake system (Table 6). The enumeration

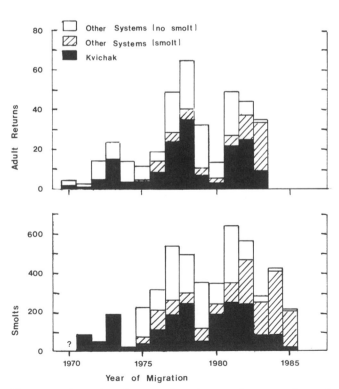

Figure 4. Annual estimates of sockeye salmon smolts and adult returns by year of seaward migration (numbers in millions of fish).

Table 3. Mean lengths of sockeye salmon smolts from Bristol Bay lake systems.
(In mm)

	1982	1983	1984	1985	1955-81 Mean	1955-81 Range
Age 1						
Kvichak	84	80	90	85	89	80, 98
Naknek	94	94	97	96	100	91,113
Egegik	104	102	106	106	—	— —
Ugashik	88[a]	89	87	93	92	81, 97
Wood	78	86	92	91	82	69, 88
Nuyakuk	—	79	81	85	—	— —
Age 2						
Kvichak	103	98	104	102	110	97,122
Naknek	100	110	108	109	112	105,120
Egegik	130	117	112	123	—	— —
Ugashik	113[a]	111	102	107	114	104,125
Wood	98	98	97	91	99	88,114
Nuyakuk	—	92	93	94	—	— —
Combined						
Kvichak	91	97	96	86	98	84,110
Naknek	95	108	104	105	107	92,116
Egegik	110	115	110	109	—	— —
Ugashik	95[a]	95	95	102	102	94,114
Wood	81	88	92	91	84	74, 89
Nuyakuk	—	80	83	85	—	— —

[a] From limited sampling (June 6-8).

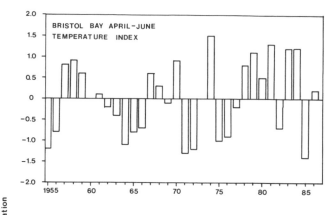

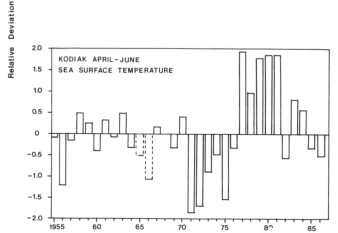

Figure 5. Annual relative deviations from mean spring temperatures in Bristol Bay and spring sea surface temperatures at Kodiak, 1955-86.

Table 4. Quarterly means of surface temperature at Kodiak, 1951-86.

	(In °C)			
	Jan-Mar	Apr-Jun	Jul-Sep	Oct-Dec
1951	0.9	6.6	12.5	4.5
1952	0.5	6.0	11.1	4.7
1953	1.3	6.6	11.9	4.4
1954	0.3	6.5	12.4	4.3
1955	0.9	6.5	11.5	3.2
1956	0.8	5.1	10.7	1.7
1957	0.4	6.4	11.9	4.2
1958	1.8	7.2	10.2	3.2
1959	0.4	6.9	11.2	4.2
1960	1.3	6.1	10.7	4.2
1961	1.3	7.0	10.7	2.5
1962	0.2	6.5	11.7	4.5
1963	2.1	7.2	13.1	4.4
1964	1.4	6.2	10.9	3.2
1965	0.2	6.0[a]	11.1[a]	4.2[a]
1966	1.6[a]	5.2[a]	9.3[a]	3.8[a]
1967	1.2[a]	6.8	11.1	5.1
1968	1.6	6.6	10.8	5.1
1969	1.6	6.2	10.9	5.6
1970	3.5	7.1	10.2	4.8
1971	0.3	4.3	9.8	4.1
1972	0.2	4.5	9.0	4.9
1973	1.9	5.5	9.3	4.7
1974	1.4	6.0	10.7	5.2
1975	1.5	4.7	10.1	5.3
1976	2.5	6.2	11.3	6.7
1977	5.7	9.0	12.5	6.1
1978	4.3	7.8	12.2	6.8
1979	4.2	8.8	13.2	6.9
1980	4.2	8.9	13.0	6.7
1981	5.5	8.9	12.4	5.2
1982	2.1	5.9	10.4	4.7
1983	3.3	7.6	11.5	6.0
1984	3.3	7.3	11.9	5.8
1985	3.3	6.2	11.2	5.1
1986	2.8	6.0	10.7	5.2

Source: Calculated from daily observations available from U.S. Department Commerce, NOAA, National Ocean Survey.

[a] Estimates from offshore temperatures; see text.

projects usually began before the migration was under way but terminated when smolts were still leaving, although usually in relatively small numbers. The beginning of the migration was thus better known than the end. In Wood River the sonar enumeration beginning in 1975 extended into August, whereas the fyke-net sampling in earlier years generally terminated about July 25. This mainly affected the dates by which 50 percent and 90 percent of the migration were reported to have occurred.

The beginnings of the Kvichak migrations since 1971 (data collected by sonar enumeration) were well correlated with the spring temperature index ($r = -.88$, $n = 14$), and the correlation for earlier years (fyke-net sampling) was also significant ($r = -.66$, $n = 14$). About 80 percent of the Kvichak smolts in the annual migrations since 1971 migrated to sea during a 4- to 20-day period. Poor marine survival (0.6 percent to 7.5 percent) was associated with both the longest migration (20 days

in 1980) and the shortest migrations (4 days in 1976 and 5 days in 1971). In other years, 80 percent of the migration occurred in 8 to 16 days and marine survival ranged from 7.6 percent to 14.1 percent.

The timing of the smolt migrations from the Wood River lake system is influenced by the numbers of juveniles in each of the five lakes. The earliest smolts to leave are from the lower lake, Lake Aleknagik, which is free of ice about 2 weeks before the uppermost lake. Thus the migrations

Table 5A. *Averages of daily water temperature in the Kvichak and Wood rivers.*

Year	Kvichak River		Wood River		
	May 17-31	June 1-15	June 1-15	June 16-30	July 1-15
1969	2.5	6.1	3.9	5.9	11.2
1970	7.6	10.3	6.4	8.1	10.2
1971	1.7	2.6	3.9	5.4	11.5
1972	1.1	4.0	3.5	4.4	7.2
1973	4.0	6.7	4.7	7.5	9.0
1974	5.7	8.0	10.9	10.7	14.2
1975	3.8	4.4	3.2	4.8	5.8
1976	2.2	4.3	4.0	5.4	7.8
1977	4.9	7.6	4.2	6.6	9.4
1978	6.6	9.1	6.0	6.9	8.0
1979	—	8.6	5.5	7.2	10.9
1980	4.5	7.0	5.2	6.4	8.2
1981	8.2	9.1	7.3	12.1	12.7
1982	3.8	6.4	4.1	5.1	5.5
1983	6.6	9.3	6.8	8.7	10.7
1984	6.7	9.2	9.0	9.7	14.2
1985	2.4	3.9	3.3	4.9	6.6
1955-1985					
Mean	4.1	6.6	5.1	7.2	9.7
S.D.	2.1	2.4	1.8	1.9	2.6
Range	0.8	1.8	3.0	4.4	5.5
	8.2	10.3	10.9	12.1	14.5

Table 5B. *Nearshore water temperatures along the North Peninsula.*

		Depth (m)	Temperature (°C)
1969	6/17-21	10	5.5
1970	6/26-7/4	20	7.5
1971	6/10-20	10	2.5
1972	June	30	3.0
1973	—	—	—
1974	June	30	4.0
1975	June	30	1.0
1976	June	30	1.0
1977	June	30	3.0
1978	June	30	5.0
1979	June	30	6.0
1980	—	—	—
1981	June	30	8.0
1982	June	30	1.0
1983	—	—	—
1984	6/30-7/3	0	9.0
1985	6/26-7/3	0	7.0

Source: 1969-71, Straty and Jaenicke (1980); 1972-82 (bottom temperatures, ca. 30 m), J. Ingraham, NMFS, Seattle, WA; 1984-85, Isakson et al. (1986).

Note: Data are from North Peninsula between Port Heiden and Port Moller.

from the lake system are quite prolonged. Since 1975, 80 percent of the migration occurred during 29 to 57 days. The onset of the migration (date of 10 percent) was poorly correlated with the spring temperature index ($r = -.58$, $n = 11$); however, the 1980 migration was unusually late relative to the warm temperature deviation that year. Excluding 1980, the correlation between timing and temperature was much higher ($r = -.85$, $n = 10$). Marine survival was also low for the 1980 smolts (4.9 percent compared to a mean of 5.7 percent and range of 3.8 percent to 9.8 percent) as was the case for the Kvichak stocks.

The timing of all the major Bristol Bay sockeye stocks can be compared for the 1982-85 migrations. Fortunately, two of these years were warm (1983-84) and the other two were cold (see Figure 5). The timing of the majority of sockeye smolts

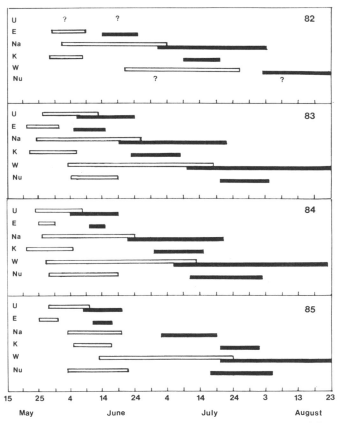

Figure 6. *Timing of sockeye salmon smolt migrations at lake system outlets (white bars) and the projected timing past Port Heiden (black bars).*

Notes: Lengths of bars denote dates of 10 percent to 90 percent of the migrations through August 23. Ugashik (U), Egegik (E), Naknek (Na), Kvichack (K), Wood River (W) and Nuyakuk (Nu), 1982-85.

Table 6. Timing of smolt migrations from Bristol Bay river systems.

	Kvichak			Ugashik			Naknek			Wood River		
	10%	50%	90%	10%	50%	90%	10%	50%	90%	10%	50%	90%
1951										6/7	6/23	7/11
1952										6/12	6/25	7/18
1953										6/3	6/17	6/23
1954										6/2	6/10	6/15
1955	6/5	6/5	6/8	–	–	–	–	–	–	6/26	7/10	7/15
1956	6/1	6/5	6/15	–	–	–	–	–	–	6/16	7/6	7/12
1957	5/31	6/1	6/24	–	–	–	–	–	–	6/11	6/24	6/26
1958	5/22	5/27	6/13	5/23	5/29	6/5	5/28	6/21	7/7	6/9	6/15	7/1
1959	5/26	5/30	6/1	5/29	5/31	6/15	6/3	6/17	7/10	6/6	6/18	6/25
1960[a]	–	–	–	6/2	6/5	6/12	6/4	6/13	6/25	6/2	6/18	7/10
1961[2]	–	–	–	5/16	5/28	6/20	6/6	6/14	7/1	6/5	6/15	7/2
1962	6/2	6/9	6/15	5/16	5/30	6/9	6/2	6/8	6/18	6/13	6/21	7/5
1963	5/25	5/27	6/7	5/16	5/31	6/10	6/1	6/19	7/1	6/9	6/16	7/2
1964	6/4	6/7	6/13	5/25	6/5	6/9	6/9	6/16	7/2	6/21	6/30	7/5
1965	5/24	5/26	5/29	5/27	6/3	6/13	6/3	6/15	6/27	6/18	7/1	7/11
1966	6/5	6/7	6/11	–	–	–	6/6	6/14	6/22	6/17	6/26	7/8
1967	5/26	6/1	6/9	5/23	5/28	6/8	5/31	6/8	6/26	–	–	–
1968	5/21	5/23	5/27	5/23	5/27	6/5	6/3	6/8	6/26	–	–	–
1969	5/28	6/1	6/12	5/25	5/30	6/5	6/4	6/9	6/29	–	–	–
1970	5/22	5/27	6/3	5/19	5/29	6/6	6/5	6/6	6/26	–	–	–
1971	6/10	6/10	6/15	–	–	–	6/9	6/13	6/25	–	–	–
1972	6/8	6/12	6/17	5/28	6/12	6/18	6/9	6/11	6/20	–	–	–
1973	5/23	5/25	5/31	5/27	5/29	6/4	5/28	6/3	6/13	–	–	–
1974	5/23	5/27	6/1	5/27	5/29	6/7	5/31	6/3	6/21	–	–	–
1975[a]	–	–	–	–	–	–	6/6	6/9	6/27	6/14	7/2	7/13
1976	6/9	6/11	6/13	–	–	–	6/4	6/7	6/16	6/20	7/14	7/29
1977	5/25	6/1	6/10	–	–	–	6/3	6/14	6/22	6/12	6/25	7/28
1978	5/20	5/24	5/29	–	–	–	–	–	–	6/7	7/3	7/24
1979	5/21	5/26	6/4	–	–	–	–	–	–	6/7	6/30	7/25
1980	5/22	5/28	6/11	–	–	–	–	–	–	6/29	7/10	7/31
1981	5/22	5/25	5/31	–	–	–	–	–	–	6/5	7/1	8/1
1982	5/28	5/31	6/8	–	–	–	6/1	6/11	7/4	6/21	7/3	7/26
1983	5/22	5/26	6/6	5/26	6/1	6/13	5/24	6/9	6/26	6/3	6/28	7/18
1984	5/21	5/27	6/5	5/24	5/31	6/8	5/26	6/6	6/24	5/31	6/24	7/13
1985	6/5	6/9	6/17	5/29	6/4	6/10	6/3	6/8	6/20	6/13	6/29	7/24

Note: Dates are those by which 10 percent, 50 percent, and 90 percent of enumerated smolts had migrated past the lake outlet.
[a] Ice flow in the Kvichak River negated accurate estimates but the migration in 1975 was in June.

leaving each of the lake systems (dates of 10 percent to 90 percent of the migrations) and the projected timing of the migrations past Port Heiden are shown in Figure 6. The timing at Port Heiden (migration out of the inner bay) was estimated from the timing at enumeration sites, the distances to Port Heiden, the mean lengths in the migrations from each lake system, and an assumed swimming speed of one body length per second (bls). Fish can maintain swimming speeds up to about 2 bls (Beamish 1978); however, Bax (1985) estimated 0.9 bls from rather imprecise mark-recapture data presented by Straty (1974), though without considering possible delay from the effects of marking.

The majority of sockeye from the Egegik and Ugashik stocks did not overlap the majority of smolts from the Naknek and Kvichak stocks and were well separated from the Nushagak stocks. The degree of separation appeared to be greater in cold years than in warm years, mainly because there was a greater delay in the Kvichak and Wood River migrations in the cold years (1982 and 1985). The simulated daily numbers of sockeye salmon smolts that migrated past Port Heiden (Figure 7) were quite variable within and between the years; however, mortalities would reduce the numbers at Port Heiden and variation in the body size of individuals would tend to spread the distributions of the stocks.

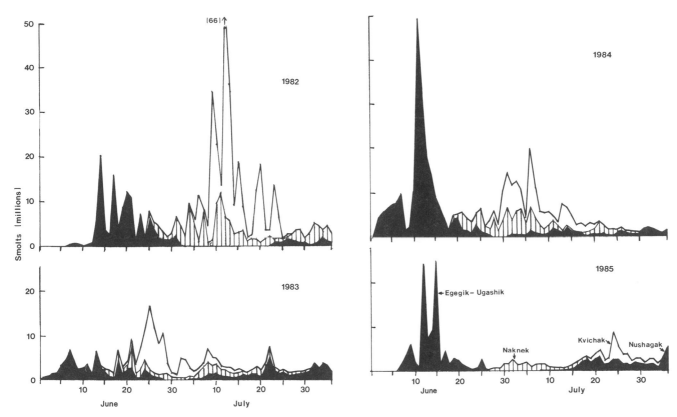

Figure 7. Simulated daily abundances of sockeye salmon smolts past Port Heiden through August 5, 1982-85.

An average Ugashik or Egegik 2-year-old smolt (11 cm) could reach the outer boundary of Bristol Bay at Cape Sarichef (about 345 km from Port Heiden) in about 36 days, arriving there by mid-July, whereas an average Wood River 1-year-old smolt (8 cm) would not arrive until 10 weeks later, at the end of September. However, the migratory rates of the smolts probably decrease in the vicinity of Port Moller, where the fish may spend more time feeding (Straty and Jaenicke 1980) and the stocks may become more mixed than they were in the inner bay.

TEMPERATURE AND GROWTH

Interannual variation in temperature may affect not only the timing of the smolt migrations from the lake systems but also the rate of migration, growth rate, and the length of time for growth during the smolts' first summer at sea. The annual mean weights of immature sockeye salmon (ages

1.1 and 2.1) from purse-seine sampling at Adak during 1958-77 were correlated with mean water temperature in Bristol Bay rivers in the preceding year ($r = .55$ and .64; Rogers 1980). Warm temperatures were associated with larger fish and cold temperatures with smaller fish.

Year-to-year variation in (1) body size of seaward migrants from Bristol Bay rivers; (2) the relative abundance of the stocks; and (3) the abundance of non-Bristol Bay stocks south of Adak could contribute considerable variation to our estimates of the mean body size of Bristol Bay sockeye salmon after 1 year at sea. The mean lengths of immature sockeye south of Adak on July 21 were still correlated with the spring temperature index in the preceding year for observations collected through 1986 (Table 7 and Figure 5). The linear correlations were .6 for age 1.1 and .58 for age 2.1 ($n = 29$). Thus spring temperatures explained 44 percent and 34 percent of the interannual variation in mean lengths; however, it was shown that temperature and migratory timing are correlated.

Table 7. Mean lengths of immature sockeye salmon south of Adak on July 21, 1958-86, and the adult returns to Bristol Bay.

	Length (mm)		Adult return (In millions of fish)
	Age 1.1	Age 2.1	
1958	337	362	15
1959	341	358	44
1960	322	352	9
1961	328	366	10
1962	340	361	6
1963	331	353	13
1964	311	335	62
1965	331	354	8
1966	323	355	10
1967	333	347	7
1968	332	353	20
1969	335	359	45
1970	320	352	9
1971	356	375	5
1972	300	334	2
1973	334	350	14
1974	341	355	23
1975	343	370	13
1976	331	366	12
1977	338	352	19
1978	333	364	49
1979[a]	347	368	65
1980	334	354	32
1981	329	340	13
1982	357	371	49
1983	325	341	44
1984	357	369	35
1985	341	356	10+
1986	316	341	—

[a] Mean lengths since 1979 were estimated from gillnet samples (Takagi and Ito 1986) and adjusted to 21 July based on the average date of catches and a growth rate of 1 mm/day. Mean lengths also adjusted to earlier means from purse-seine sampling by subtracting 6 mm (mean difference between means from purse-seine samples and gillnet samples during simultaneous sampling, 1972-78).

The timing of the Kvichak smolt migrations since 1971 (when sonar estimates were available) explained nearly as much of the interannual variation in mean lengths of immature sockeye salmon as did temperatures during the same years (Figure 8). When migrations occurred in May (90 percent by 31 May) the mean lengths of age 2.1 and 1.1 sockeye salmon at Adak were about 2 cm longer than when migrations were mainly in June (90 percent by 15 June, day 46).

Spring weather in 1985 was cold and smolt migrations were late; however, water temperatures in Bristol Bay were relatively warm, providing a unique situation. If interannual variation in water temperatures largely affected the growth rate of seaward migrants, then we would expect the immature sockeye at Adak in 1986 to be at least average in size. Instead, they were well below average (see Table 7). Spring temperatures probably affect timing of seaward migration more than growth. Timing probably affects the period of time the fish have to grow and thus their size a year later.

There was no significant correlation nor partial correlation with abundance as measured by the number of returning adults; therefore, there was no indication that body size of Bristol Bay sockeye salmon after 1 year at sea was density dependent. This was somewhat unexpected because smolts at a given age tend to be smaller in years of high abundance (Rogers 1980). The spatial separation of the stocks in their seaward migration and the possibility of size-dependent marine survival may negate density effects on growth.

MARINE SURVIVAL

The interannual variation in marine survival of Bristol Bay stocks can only be examined for the Kvichak and Wood River sockeye salmon and the time series are very short (12 years and 9 years, see Table 2). Marine survival was somewhat correlated for the two stocks for the 8 years since 1976,

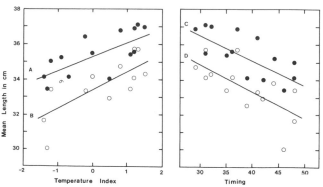

Figure 8. Mean lengths of immature sockeye salmon south of Adak (1972-1986, excluding 1976) regressed on temperature index (left) and timing (right) of Kvichak smolt migration (date of 90 percent, May 1 = 1).
A. Age 2.1 $Y = 35.3 + 0.84(X_1)$, $r = .73$
B. Age 1.1 $Y = 33.4 + 1.04(X_1)$, $r = .74$
C. Age 2.1 $Y = 40.7 - 0.14(X_2)$, $r = .71$
D. Age 1.1 $Y = 39.7 - 0.16(X_2)$, $r = .69$

but the correlation was not statistically significant (r = .52); and there was no indication of density-dependent survival for either stock.

Several factors were somewhat associated with marine survival for the Kvichak sockeye salmon smolts; body size, freshwater age composition (which was also correlated), temperature, timing of migration (which was also correlated), and ocean age composition in the adult returns. The correlation with the ocean age composition was the only statistically significant correlation (r = -.86, n = 12); however, ocean age was also variously correlated with the other factors (with the exception of timing of seaward migration). Kvichak sockeye tend to return after 2 years at sea; thus relatively high proportions of older fish are associated with low returns of adults.

The spring temperature index was the only factor associated with marine survival of Wood River sockeye salmon and the correlation (r = .64) was not quite significant at α = .05, but was higher than the correlation for the Kvichak stock (r = .40). Age composition and body size did not vary from year to year as much for the Wood River sockeye salmon as for the Kvichak sockeye salmon. The Wood River stocks tended to produce more smolts per spawner than the Kvichak stocks,[1] but at a smaller size and lower marine survival.[2] The relative production (return per spawner) was about the same for the two stocks.[3]

Some conditions at the time of seaward migration (large body size, warm temperatures, early migration) appear to contribute to better marine survival for Bristol Bay sockeye salmon; however, it is not clear whether these conditions contribute to better survival during the first few months at sea (in Bristol Bay) or at a later time and place. Eggers et al. (1984) showed that the relative survival (run per average of parent escapements) of Bristol Bay sockeye salmon was correlated with the mean of June air temperatures at Cold Bay (near the tip of the Alaska peninsula) in the 2 years prior to the year of the run. June air temperatures at Cold Bay were correlated with the spring temperature indices (r = .64) and the April-June water temperatures at Kodiak (r = .72) shown in Figure 5.

Stepwise multiple regression analysis was used to examine possible sources of variation in marine survival for the Kvichak stocks (see Table 2). Mean weight of smolts, temperature index, and date of 90 percent of the migration were used as measures of conditions at time of seaward migration. Annual (July-June) water temperatures at Kodiak during the first year and the second year at sea were used as measures of the physical environment, and the mean lengths of immature sockeye salmon (unweighted mean of age 1.1 and 2.1 from Table 7) were used as a measure of relative growth after 1 year at sea.

The mean length of immature sockeye salmon at Adak was the first variable to enter the analysis (r = .68) and temperatures at Kodiak during the following year (second year at sea), which was also correlated with survival (r = .65), was the second variable to enter (R = .82, F = 9.5, d.f. 2, 9). Kodiak temperatures during the first year at sea (correlated with temperatures in the second year, r = .84) also entered the analysis (R = .92); however, the coefficient was negative and the partial correlation was due almost entirely to one observation (1980). This analysis merely indicates that factors associated with conditions beyond the estuary may cause as much interannual variation as conditions at seaward migration. The utility of the various models to predict marine survival was rather poor; e.g., for the 1984 Kvichak migration of eighty-nine million survival was predicted at 7 to 9 percent, yet with the poor return of two-ocean fish in 1986 of just over one million, it is very unlikely that marine survival will exceed 4 percent after the return of three-ocean fish in 1987.

Information gathered to date on Bristol Bay sockeye salmon suggests that growth and mortality in marine waters are largely independent of density. However, we are still unable to predict marine survival with accuracy.

Some improvement in our ability to do so for Bristol Bay salmon may come when a longer time series is available; however, real progress depends on determining the causes of mortality and measurement of biological and environmental parameters at the time and place where the fish occur during their ocean life. There is an obvious need for marine research on Pacific salmon and yet our current efforts are very inadequate.

---

[1] 1974-81 brood years, W.R. $\bar{X}$ = 50, S.D. = 32, Kvi. $\bar{X}$ = 31, S.D. = 13.

[2] W.R. $\bar{X}$ = 6%, S.D. = 2.3, Kvi. $\bar{X}$ = 9%, S.D. = 4.8.

[3] W.R. $\bar{X}$ = 3.0, S.D. = 1.5, Kvi. $\bar{X}$ = 2.8, S.D. = 1.3.

LITERATURE CITED

Bax, N.J. 1985. Simulation of the effects of oil spill scenarios on juvenile and adult sockeye salmon *(Oncorhynchus nerka)* migrating through Bristol Bay, Alaska. NOAA NMFS, NWAFC Processed Rept. 85-03. 128 p.

Beamish, F.W.H. 1978. Swimming capacity. Pages 101-187 in W.S. Hoar and P.J. Randall, eds., *Fish Physiology*. Vol. VII. Locomotion. Academic Press.

Brett, J.R., and N.R. Glass. 1973. Metabolic rates and critical swimming speeds of sockeye salmon *(Oncorhynchus nerka)* in relation to size and temperature. *J. Fish. Res. Board Can.* 30:379-387.

Eggers, D.M., and D.E. Rogers. 1978. Modeling the migration of sockeye salmon smolts through Bristol Bay, Alaska. Fisheries Research Inst., Univ. Washington. Final Rep. FRI-UW-7806. 95 p.

Eggers, D.M., and H.J. Yuen, eds., 1984. 1982 Bristol Bay sockeye salmon smolt studies. ADF&G Tech. Data Rep. 103. Juneau. 72 p.

Eggers, D.M., C.P. Meachum, and D.C. Huttunen. 1984. Population dynamics of Bristol Bay sockeye salmon, 1956-1983. Pages 200-225 in W.G. Pearcy, ed., *The Influence of Ocean Conditions on the Production of Salmonids in the North Pacific*. Oregon State University Sea Grant College Program, Corvallis. (ORESU-W-83-001.)

French, R., H. Bilton, M. Osako, and A. Hartt. 1976. Distribution and origin of sockeye salmon *(Oncorhynchus nerka)* in offshore waters of the North Pacific Ocean. *Int. N. Pac. Fish. Comm., Bull.* 34. 113 p.

Frost, K.J., L.F. Lowry, and R.R. Nelson. 1983. Belukha whale studies in Bristol Bay, Alaska. Pages 187-200 in *Proceedings of the Workshop on Biological Interactions among Marine Mammals and Commercial Fisheries in the Southeastern Bering Sea*. Oct. 18-21, 1983, Anchorage. Alaska Sea Grant Rep. 84-1.

Harris, C.K., and D.E. Rogers. 1979. Forecast of the sockeye salmon run to Bristol Bay in 1979. Fisheries Research Inst., Univ. Washington. Circ. 79-2. 50 p.

Hartt, A.C. 1980. Juvenile salmonids in the oceanic ecosystem. Pages 25-57 in W.J. McNeil and D.C. Himsworth, eds., *Salmonid Ecosystems of the North Pacific*. Oregon State University Press, Corvallis.

Isakson, J.S., J.P. Houghton, D.E. Rogers, and S.S. Parker. 1986. Fish use of inshore habitats north of the Alaska Peninsula June-September 1984 and June-July 1985. Final Rep. to MMS and NOAA. Dames and Moore and Univ. Washington, Seattle. 236 p.

Mathews, S.B. 1984. Variability of marine survival of Pacific salmonids: a review. Pages 161-182 in W.G. Pearcy, ed., *The Influence of Ocean Conditions on the Production of Salmonids in the North Pacific*. Oregon State University Sea Grant College Program, Corvallis. (ORESU-W-83-001.)

Mesiar, D.C. 1986. Timing, distribution, and estimated abundance of seaward migrating pink salmon *(Oncorhynchus gorbuscha)* in the Nushagak River, Alaska. M.S. Thesis, Univ. Alaska, Juneau. 88 p.

Ricker, W.E. 1976. Review of the rate of growth and mortality of Pacific salmon in salt water, and noncatch mortality caused by fishing. *J. Fish. Res. Board Can.* 33(7):1483-1524.

Rogers, D.E. 1978. Determination and description of knowledge of the distribution, abundance, and timing of salmonids in the Gulf of Alaska and Bering Sea, a supplement to the final report. NOAA/OCSEAP RU #353, Final Rep. Biol. Studies 1:205-235.

Rogers, D.E. 1980. Density-dependent growth of Bristol Bay sockeye salmon. Pages 267-283 in W.J. McNeil and D.C. Himsworth, eds., *Salmonid Ecosystems of the North Pacific*. Oregon State University Press, Corvallis.

Rogers, D.E. 1984. Trends in abundance of Northeastern Pacific stocks of salmon. Pages 100-127 in W.G. Pearcy, ed., *The Influence of Ocean Conditions on the Production of Salmonids in the North Pacific*. Oregon State University Sea Grant College Program, Corvallis. (ORESU-W-83-001.)

Rogers, D.E. (In press). Pacific salmon. Chapter 13. In D.W. Hood and S.T. Zimmerman, eds., *The Gulf of Alaska*. NOAA, U.S. Dept. Commerce.

Schumacher, J.D., and P.D. Moen. 1983. Circulation and hydrography of Umimak Pass and the shelf waters north of the Alaska Peninsula. NOAA Tech. Memo. EREL PMEL-47. Pacific Marine Environmental Laboratory. Seattle. 75 p.

Shaul, A.R., L.J. Schwarz, J.N. McCullough, and L.M. Malloy. 1985. Alaska Peninsula-Aleutian Islands area annual finfish management report. ADF&G Finfish Data Rep. No. 1-86. Kodiak. 237 p.

Straty, R.R. 1974. Ecology and behavior of juvenile sockeye salmon *(Oncorhynchus nerka)* in Bristol Bay and the eastern Bering Sea. Pages 285-319 in D.W. Hood and E.J. Kelly, eds., *Oceanography of the Bering Sea*. Occasional Publ. No. 2. IMS, Univ. Alaska, Fairbanks.

Straty, R.R. 1975. Migratory routes of adult sockeye salmon *(Oncorhynchus nerka)* in the Eastern Bering Sea and Bristol Bay. NOAA Tech. Rep. NMFS SSRF-690. 32 p.

Straty, R.R., and H.W. Jaenicke. 1980. Estuarine influence of salinity, temperature, and food on the behavior, growth, and dynamics of Bristol Bay sockeye salmon. Pages 247-265 in W.J. McNeil and D.C. Himsworth, eds., *Salmonid Ecosystems of the North Pacific*. Oregon State University Press, Corvallis.

Takagi, K., and S. Ito. 1986. Abundance and biological information of immature sockeye salmon in waters south of the Aleutian Islands in July, 1986. (Document submitted to annual meeting of the Int. N. Pac. Fish. Comm., Anchorage, AK, Oct. 1986). Fisheries Agency of Japan. 31 p.

Yuen, H.J., M.L. Nelson, and R.E. Minard. 1986. Bristol Bay salmon (*Oncorhynchus* spp.)—1982, a compilation of catch, escapement, and biological data. ADF&G Tech. Data Rep. 175. Juneau. 105 p.

Problems of Managing Mixed-Stock Salmon Fisheries

Donald E. Bevan

Professor Emeritus
School of Fisheries
University of Washington

The problems of managing mixed-stock or mixed-species fisheries were recognized very early in the history of salmon fisheries. Rathbun (1899), in a review of an international commission's work from 1893 to 1896, said:

> In the region to which this paper relates there may still be time to give the fisheries the full benefits of a wise protection before any of its branches shall have been appreciably impaired, but action should not be long deferred, as a decrease once begun is hard to check . . . Any system of protective regulations should therefore contemplate for the welfare of the entire salmon group; but with some species there is much greater urgency for action than with others.

Many authors have since recognized the problem, but few have offered solutions. Henry et al. (1986), in a review of 1985 Pacific Coast ocean salmon fisheries off California, Oregon, and Washington, pointed out a number of problems with the management of mixed-stock and mixed-species fisheries. A major problem was the lack of overall season quotas resulting in ocean catches that were greater than anticipated and reduced returns to fisheries and for escapements. It was found that quotas had been inappropriately applied to the

1985 ocean fishery because there was no practical way to control catch when the quota was taken in less than one day. Also, because the amount of effort was difficult to predict and control, the existing monitoring system was incapable of effectively constraining catches within the established quota.

I suggest it is useful before proceeding further to define the objectives of managing a mixed-stock salmon fishery. The primary objective should be the control of fishing effort to provide sufficient escapement for each of the stocks so that return for the sum of the stocks is optimized. Clearly, the optimum cannot be the maximum yield for each stock, unless each stock is fished separately. While a greater yield might be possible by abandoning mixed-stock ocean fisheries and fishing only in terminal areas where stock intermingling is reduced, I doubt that the economic, social, and political costs of such a policy would justify the gains.

I purposely use the term "maximum yield," not "maximum sustained yield." Salmon fisheries commonly exhibit pronounced regular natural cycles and irregular natural fluctuations. This causes me to view the word "sustained" as having little value in describing or modeling a salmon fishery. The optimum may not produce the maximum from any stock. Its value will depend upon decisions regarding allocations to different fisheries and the relative values of short- and long-term returns.

In 1976 a group of distinguished scientists met in Tiburon, California, to define the problems of mixed-stock fisheries and to identify the types of research that could be expected to best benefit fishery management (Hobson and Lenarz 1977). Although very few of the recommendations have been carried out (Abramson 1986), the report is useful in defining the problem. The group recognized that most mixed-stock fisheries have been managed as if the key elements were independent; they identified five interacting elements: (1) management objectives; (2) regulations; (3) fleet or fishers; (4) exploited stocks; and (5) ecosystems. The group recognized that the continuously changing nature of these elements vastly increased the complexity of the interactions among them. A matrix was developed to describe variables that characterize a fishery. Figure 1 illustrates the increasing complexity of managing fisheries as intermingling of stocks increases and as fishing gear becomes less selective for target stocks. Figure 1 provides little confidence for management options in many mixed-

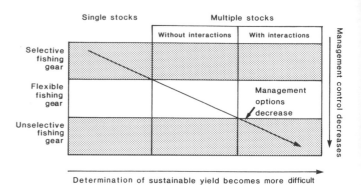

Figure 1. *Complexity of management as a function of selectivity of fishing gear and intermingling of stocks.*
Adapted from Hobson and Lenarz 1977.

stock salmon fisheries, since only biological interactions are considered. Biological interactions in salmon are either unknown or unclear, and we have to take into account the complicating factor of technological interaction (Murawski 1984).

Before considering models for mixed-stock management, it may be useful to consider the bases for single-stock management. At the heart of the process is the setting of an escapement goal. Royce (1964) called this the optimum escapement. He estimated that there were 10,000 genetically different salmon stocks in the Pacific Northwest and Alaska. Optimum escapement can differ considerably among stocks, depending upon the productivity of fresh- and saltwater ecosystems and the effects of density-dependent mechanisms on stocks. With some controlling assumptions, various authors have proposed models to estimate optimum escapement for individual stocks (Ricker 1954, 1958, Beverton and Holt 1957, Paulik and Greenough 1966).

These models typically describe the number of adults returning (recruits) as a function of the number of parents. Various parameters are introduced to equations to insure that the spawner-recruit relationship is described by a dome-shaped curve. Therefore, the number of spawners producing the largest number of recruits is intermediate between low and high escapement. Escapement goals are frequently obtained from such models.

With an escapement goal established, management regulations are promulgated to control fishing effort. This process is fraught with difficulty since a reliable forecast of return abundance is necessary to develop a catch quota as a means of control. Given the high variability in return from similar spawning stocks and the uncertainty of forecasts of run size, many authors have suggested that a better approach toward obtaining optimum escapement is to control fishing effort in such a way as to obtain a constant percentage escaping rather than a fixed number. Quinn et al. (1985), in reference to Pacific halibut, point out that quotas set to provide maximum sustainable yield are difficult to use in actual practice. They may be correct over a long term, but annual quotas may result in overexploitation due to fluctuations in stock size. Instead, these authors suggest taking a fixed percentage of the stock each year. They call this fixed percentage "constant exploitation yield." It is defined as the amount of yield obtained by taking catches proportional to the stock abundance. The proportionality constant is calculated to produce maximum sustainable yield when the stock biomass is at the appropriate level.

I have attempted to find both the origin of the idea of a fixed percentage of harvest for salmon management and its original scientific justification. The idea seems first to have been proposed in 1924 in the White Act (Bower 1925a, b). The Act states:

> In all creeks, streams, or rivers or in any other bodies of water in Alaska, over which the United States has jurisdiction, in which salmon run, and in which now or hereafter there exists racks, gateways, or other means by which the number in the run may be counted or estimated with substantial accuracy, there shall be an escapement of not less than 50 per centum of the total number thereof. In such waters the taking of more than 50 per centum of the run of such fish is hereby prohibited. It is hereby declared to be the intent and policy of Congress that in all waters of Alaska in which salmon run there shall be an escapement of not less than 50 per centum of the total number thereof, and in any year it shall appear to the Secretary of Commerce that the run of fish in any waters has diminished, or is diminishing, there shall be required a correspondingly increased escapement of fish therefrom.

The White Act had been proposed unsuccessfully in the Congress for many years before it was adopted in 1924. I have gone through many pages of testimony before the House Merchant Marine and Fisheries Committee, and the justification for the 50 percent escapement originated with the catch and weir counts of red salmon from Karluk, Kodiak Island (Gilbert 1924). Gilbert planned to experiment with several streams and various percentage escapements, and he states:

> In the Karluk River, which was the first of the Alaska streams we subjected to this experiment, in the first year we found that approximately 50 percent of the entire run was being captured by the fishermen and 50 percent, approximately, escaped from them and ascended the river to the spawning grounds.

I now believe that the application of a rate of harvest is a more feasible management policy than an attempt to obtain a fixed escapement each year. However, the rate should be adjusted, depending on size of stocks. More protection from exploitation is needed at low population levels and less protection at high population levels. This is not only self-evident but has been borne out by stock recruitment analyses that show that productivity is underestimated at high stock levels and overestimated at low stock levels (Walters 1985, Hilborn 1985). I also concur with the ideas of Walters and Hilborn (1976) and Tyler et al. (1982) that alternatives should be tested as deliberate experiments.

It is tempting to be pessimistic and adopt the view of McDonald (1981) that, with the large number of stocks involved and with intensive fishing on mixed stocks, the development of useful stock-recruitment relationships as a basis for setting escapement goals is precluded. While I concur with his view that it is difficult to be optimistic about all but the most abundant stocks, I believe on proceeding on the basis of a suggestion made verbally by Carl Walters which I will try and paraphrase: "If you don't know what you're doing, don't just sit there—do something—at least you should gather some more information."

I suggest that the application of a quota to provide a fixed escapement has limited utility in a mixed-stock fishery. An overall quota that will achieve the desired escapement of the weakest stock will result in underfishing of all stronger stocks. On the other extreme a quota which will fully harvest the strongest stock will result in overfishing of all weaker stocks.

Economic, social, and political demands force us to undertake mixed-stock fisheries. Even if this were not the case, we would have similar problems in managing terminal fisheries in many estuaries, since what might be construed as a single stock is often a mixture of stocks destined to spawn in separate tributaries (Thompson 1951).

Assume that we reject managing for the weakest or strongest stocks, but select some intermediate goal. We would then need to optimize fishing rates. A starting point might be to expand the simple equation

$$F = q \cdot f$$

where
F = fishing mortality
q = catchability coefficient
f = fishing effort

to include the necessary subscripts to sum over species, age classes, time, fisheries (space), gear, etc.

Salmon have well defined timing for migrations of different stocks. Their distribution at sea and migration paths are probably reasonably well defined as well, but they are not constant. For example, sockeye returning to the Fraser River, British Columbia, sometimes approach the Fraser from the north through Georgia Strait rather than from the west through Juan de Fuca Strait.

It is possible that in years of low abundance salmon alter their distribution at sea rather than changing patterns of density. Fishing success (i.e., catch per unit of effort) depends not only on abundance but also on distribution. It also depends upon skills of vessel operators (Hilborn 1985). It is not entirely clear how catch per unit of effort relates to total abundance, distribution of salmon, and skills of fishers; but it is possible that daily catch per unit of effort can be high in years of low stock abundance as well as in years of high stock abundance.

Although I may seem to rely on the standard fallback position of the fishery biologist—I need more data—I suggest that the most useful approach is to develop our capabilities with data base management sytems (Delibero 1986). Daily catch and effort statistics, with at least some attempt to differentiate time and space where fish are caught in the ocean, should be included. Continuation of tagging programs with coded-wire tags (Jefferts et al. 1963, Bergman 1968) is essential for evaluation of contributions of hatchery stocks. These efforts offer promise for some understanding of variations in catchability coefficients for various stocks that are included in the mixture of stocks that we wish to harvest.

While you may believe I have presented a pessimistic view of the state of mixed-stock management, as have Walters and Riddel (1986), there is room

for optimism. The major experiment on mixed-stock management is just beginning. This involves primarily chinook salmon under the U.S.-Canada Salmon Treaty. Although data are preliminary, there are indications that the stocks, including the weaker ones, are responding positively to the regulatory regimes created in response to the treaty.

LITERATURE CITED

Abramson, N.J. 1986. Personal communication.

Bergman, P.K. 1968. The effects of implanted wire tags and fin excision on the growth and survival of coho salmon. Ph.D. Dissertation. College of Fisheries, University of Washington, Seattle.

Beverton, R.J.H. and S.J. Holt. 1957. On the dynamics of exploited fish populations. *Fish. Invest.* London. Ser. 2, 19, Pp. 533.

Bower, W.T. 1925a. Alaska fishery and fur seal industries in 1923. Bureau of Fisheries Document No. 973. Washington, D.C.

Bower, W.T. 1925b. Alaska fishery and fur seal industries in 1924. Bureau of Fisheries Document No. 992. Washington, D.C.

Delibero, F.E. 1986. A statistical assessment of the use of the coded wire tag for chinook and coho studies. Ph.D. Dissertation, School of Fisheries, University of Washington, Seattle.

Gilbert, C.C. 1924. Testimony presented to the House Merchant Marine and Fisheries Committee on H.R. 2714 (68th Congress), pp. 329-336.

Henry, K., et al. 1986. Review of 1985 ocean salmon fisheries. Pacific Fisheries Management Council, Portland, Oregon.

Hilborn, R. 1985. Fleet dynamics and individual variation: why some people catch more fish than others. *Can. J. Fish. Aquat. Sci.* 42:2-13.

Hobson, E.S. and W.H. Lenarz. 1977. Report of a colloquium on the multispecies fisheries problem, June 1976. *Marine Fisheries Review*, Vol. 39, No. 9, pp. 8-13.

Jefferts, K.B., P. Bergman, and H. Fiscus. 1963. A coded wire identification system for macro-organisms. *Nature* 198(4879) pp. 460-462.

Larkin, P.A. and B. Parrish. 1970. Symposium on stock and recruitment. Rapports et Proces Verbaux Cons. Int. Explor. Mer.

Murawski, S.A. 1984. Mixed species yield-recruit analyses accounting for technological interactions. *Can. J. Fish. Aquat. Sci.* 41:897-916.

McDonald, J. 1981. The stock concept and its application to British Columbia fisheries. *Can. J. Fish. Aquat. Sci.* 38:1657-1664.

Paulik, G.J., A.S. Hourston, and P.A. Larkin. 1966. Exploitation of mutiple stocks by a common fishery. *J. Fish. Res. Bd. Can.* 24(12):2527-2537.

Paulik, G.J. and Greenough. 1966. Management anaylses for a salmon resource system. Pages 215-250 in K.E.F. Watt, ed., *Systems Analysis in Ecology*. Academic Press, New York.

Quinn, T.J., II, R.B. Deriso, and S.H. Hoag. 1985. Methods of population assessment of Pacific halibut. Scientific report IPHC.

Rathbun R. 1899. A review of the fisheries in the contiguous waters of the State of Washington and British Columbia. U.S. Fish Commission Report for 1899, pp. 251-350.

Ricker, W.E. 1954. Stock and recruitment. *J. Fish. Res. Bd. Can.* 11(5):559-623.

Ricker, W.E. 1958. Maximum sustained yield from fluctuating environments and mixed stocks. *J. Fish. Res. Bd. Can.* 15(5):991-1006.

Royce, W.F. 1964. Trends in salmon fisheries. Fisheries Research Inst., University of Washington, Seattle. Circular No. 210, pp. 29-35.

Thompson, W.F. 1951. An outline for salmon research in Alaska. Fisheries Research Inst., University of Washington, Seattle. Circular No. 18.

Tyler, A., W.L. Gabriel, and W.J. Overholtz. 1982. Adaptive management based on structure of fish assemblages of northern continental shelves. Pages 149-158 in M.C. Mercer, ed., *Multispecies Approaches to Fisheries Management Advice*. Spec. Publ. 59 of Can. Fish. Aquat. Sci.

Walters, C.J. and R. Hilborn. 1976. Adaptive control of fishing systems. *J. Fish. Res. Bd. Can.* 33:145-159.

Walters, C.J. 1985. Bias in the estimation of functional relationships from time series data. *Can. J. Fish. Aquat. Sci.* 42:147-149.

Walters, C.J. and B. Riddell. 1986. Multiple objectives in salmon management: The chinook sport fishery in the Strait of Georgia, B.C. *N.W. Environ. J.* 2(1).

Mixed-Stock Fisheries and the Sustainability of Enhancement Production for Chinook and Coho Salmon

Carl J. Walters

Professor of Animal Ecology
Institute of Resource Ecology
University of British Columbia

The last three decades have seen massive growth in hatchery releases of chinook and coho salmon in the Pacific Northwest, accompanied by many promises of more fish for all. Until the early 1980s there was little indication that the promises might not be kept, though there were a few suspicions that hatchery development might be partly responsible for declines in some wild stocks. When some hatchery stocks began to show poor returns (smolt to adult survival based on coded-wire tags) beginning in the mid-1970s, fingers were immediately pointed at oceanographic factors such as up-welling rates off the Oregon coast and surface temperatures off British Columbia. But the poor survival rates have persisted, and we must now begin to wonder if there is something dreadfully wrong with the overall production system: in the 1980s, we seem to be paying more and more to produce declining catches, with some hatchery stocks showing barely higher overall productivity (egg to adult survival) than could be sustained by wild stocks alone.

This paper reviews some alternative hypotheses about why hatchery productivities have been declining, and about the mechanisms that may preclude return to a production system based mainly on wild stocks should the declines continue. These hypotheses and mechanisms have been heatedly debated by salmonid biologists, and we are not making much progress towards a scientific consensus. I conclude by suggesting that progress will continue to be slow, while risks of irreversible mistakes will continue to grow, unless a commitment is made to perform some quite drastic and large-scale experiments involving alterations in the operating policies of major production hatcheries.

DECLINES IN HATCHERY PRODUCTIVITY

Four general hypotheses have been advanced to explain (or apologize for) declines in survival rates of hatchery stocks: *(1)* environmental changes; *(2)* genetic deterioration; *(3)* ecosystem responses; and *(4)* hatchery practices. This section criticizes each of these hypotheses, and comments on the data that will be needed to test them.

Environmental factors

There is a long tradition in fisheries management of seeking various environmental indices that could explain variation in productivity. Since lots of environmental data are available, especially on oceanographic conditions, the diligent searcher can generally find some good correlations. Unfortunately there is an almost equally long tradition of having the correlations break down after a few years, which should not surprise any scientist. Quite convincing correlations in relation to hatchery and wild coho salmon have recently been published by Nickelson (1986), who reviews earlier studies and looks at upwelling and temperature in the Oregon Production Area. Nickelson's data indicate that both wild and hatchery stocks have shown declining survival, with the hatchery decline perhaps more dramatic. T. Perry (Department of Fisheries and Oceans, Vancouver, personal communication) has found excellent correlations between hatchery chinook survival and sea surface temperatures in British Columbia. Walters and Riddell (1986) examine survival trends for wild chinook stocks, and do not find trends that consistently parallel the declines seen in hatchery stocks.

So at this point we have some striking correlations and conflicting evidence about whether these correlations extend to wild stocks (which would

make them a lot more convincing, though hatchery stocks may of course be more "sensitive" to environmental factors). One might suppose that these correlations will be tested simply by waiting until oceanographic conditions change, then seeing if responses occur as predicted. Indeed, a few years of positive results (such as more upwelling with higher survivals) would be quite convincing. But suppose that negative results are seen; would the environmental hypothesis be rejected? Here the answer is no, since anyone with a vested interest in explaining away the results could simply argue that inappropriate environmental indices had been used in the original correlations, and could doubtless find other indices (upwelling in different months, temperatures at other stations, etc.) for which high correlations would still be evident.

Genetic deterioration

There has been much concern about hatchery practices that result in very low effective population sizes (e.g., fertilization of many females with each male) and that may select for genotypes that perform well in the hatchery environment but not in the field after release. These concerns are valid and may explain poor performance in some older hatchery systems. However, I doubt that they can explain the survival declines that have been seen in various newer hatcheries, particularly in British Columbia, where considerable care has been taken to maintain large brood stocks and where survival declines occurred for brood years produced by returns of the first major hatchery releases (Figures 1-2; see also Figure 7 in Chinook Technical Committee 1985). If selection for fish that are good hatchery performers but poor post-release survivors has occurred in British Columbia, it must have been very rapid and dramatic.

The obvious test of genetic deterioration hypotheses is to reintroduce wild stock into hatchery systems, then see if performance improves. In essence this test is being done all the time on an informal basis, as new hatcheries are developed and as various stocks are brought into existing facilities for experimental or conservation reasons. No consistent and obvious pattern has emerged from this ongoing experience; wild fish show extremely variable performance when first introduced into hatcheries. I am not aware of any definitive and radical experiments where an entire hatchery stock has been discarded and replaced, which would allow us to decide whether it was the stock or the hatchery environment itself that had deteriorated.

Ecosystem responses

Hatchery production systems create predictable, localized superabundances of fat and stupid (supply your own technical jargon if offended by this crude terminology) juvenile fish, and this is likely to stimulate aggregations of slightly less stupid predators and perhaps even accumulation of various microscopic pathogens. These aggregations and accumulations may require several years to develop, and could occur anywhere along the post-release migration path of the salmon.

Though various point estimates of predation rates and disease incidences have been made by researchers throughout the Pacific Northwest, no one has been able to establish that any particular predator or location has a consistently high (above 20 percent mortality) impact. Most localized (by site and predator or disease type) loss estimates are quite low, in the order of a few percent. Thus if "ecosystem responses" are responsible for declining performances, these responses must involve a variety of mortality agents and locations.

Ecosystems response hypotheses will not be tested by detailed monitoring programs that attempt to catalog the losses all along a stock's migration route; like environmental indices, negative outcomes from ecological monitoring could be explained away by pointing further out to sea (where monitoring is more difficult) or to unidentified pathogens. A simple but expensive direct test would be more effective: shut down a hatchery for some years (enough to insure that even long-lived predators forget about it), then see if it performs well when suddenly started again. A less costly (but less reliable) test would be to deliberately make the hatchery fish unpredictable, by varying the release schedule over time both within and among years. I personally doubt that the less costly alternative would be effective.

Hatchery practices

A first glance at Figures 1-2 leaves the general impression of an overall downward trend in coho survivals from the British Columbian hatcheries. But, examined more closely, the coho data indicate that declines occurred in different years for the different facilities, and were less marked or absent in those facilities that had lower survivals in the first place. In other words, the data suggest that

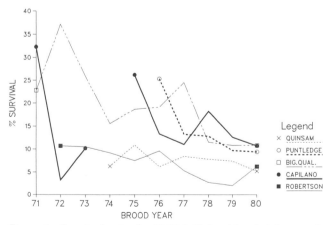

Figure 1. *Survival rates for British Columbian hatchery coho salmon.*

Notes: Estimated (catch plus escapement divided by smolt releases) from coded-wire tagging returns using a data base prepared by Carol Cross and Don Bailey, Canadian Department of Fisheries and Oceans, Vancouver. Data base did not include American catches and assumed a 30 percent "awareness factor" for voluntary tag returns by sport fishers.

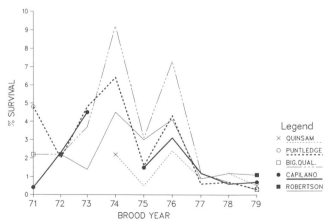

Figure 2. *Survival rates for British Columbian hatchery chinook salmon.*

Notes: Estimated (catch plus escapement divided by smolt releases) from Cross-Bailey data base as in Figure 1, with the same qualifications about American and sport catches.

each hatchery may display a characteristic "transient" response. This response may involve initial low survivals (as in Canadian chinook hatcheries, see Figure 2 and Figure 7 in Chinook Technical Committee 1985), followed in some cases by strong but temporary improvements.

These responses could be explained by improved hatchery practices followed by ecosystem responses, but a simpler hypothesis is that things go wrong within the hatchery operating system even if there are no external ecosystem responses. Obvious problems that could develop over time in large production hatcheries include accumulation of pathogenic organisms whose effects are expressed after the juveniles are released (hence no insystem detection and treatment), and deterioration in water quality and increased stress due to increasing rearing densities (as the hatchery manager discovers that increased crowding produces no obvious ill effects).

Problems with hatchery practices are likely to be associated with growth in total releases and in rearing density. It is hard to convince hatchery managers that such trends may be undesirable, but the only sure way to test the hatchery practices hypothesis is to make a substantial cut in total releases from at least two large production hatcheries. Small-scale experiments (rearing density changes, etc.) within large hatcheries may provide some useful information, but negative results from such experiments would not preclude the existence of hatchery-wide problems such as inadequate disease prophylaxis.

DECLINES IN WILD STOCKS

It may very well be discovered in the next decade that hatchery productivities cannot be economically sustained at levels high enough to justify ignoring wild stocks (the net value of a hatchery stock must be judged by how much more it can produce than the same stock spawning naturally). In view of this possibility it would be foolish to allow continued deterioration in wild stocks, at least to the degree that this deterioration is caused by hatchery production in the first place. This section reviews five basic mechanisms by which hatchery production may impact on wild stocks: (1) brood stock harvesting; (2) impaired hatchery-wild mating; (3) juvenile competition; (4) disease transmission and predator attraction; and (5) mixed-stock fishing.

Brood stock harvesting

From the standpoint of a natural population, removal of individuals for hatchery brood stock has the same impact as any other harvesting process. Hopefully the bad old days of heavy brood stock harvesting are past, but it is worth remembering this danger at a time when some hatcheries are experiencing such poor survivals that they need to look outside for egg sources.

Hatchery-wild mating

Being stupid, hatchery fish tend to stray and may then try to mate with wild fish. If the offspring from such matings have reduced survival as Chilcote et al. (1986) and others have warned, then the hatchery fish are acting like a sterile male release as used in insect control. For a depressed natural stock, even a small percentage straying rate from a much larger hatchery stock nearby may mean that a large percentage of the potential natural production is lost.

Competition among juveniles

There is little evidence of density dependence in marine survival rates of chinook and coho, except that recent overall trends in survival rates may reflect impacts of increasing total smolt abundance. However, more direct and obvious competition (for space) may occur in fresh water when hatchery fry and presmolts are outplanted; the hatchery juveniles are often larger, and may enjoy an advantage in behavioral contests for secure territories (Berg and Northcote 1985, Chapman 1962, Dill 1978, Fenderson et al 1968, Mason and Chapman 1965, Solazzi et al. 1983).

Disease transmission and predator attraction

There is often concern about wild stocks as disease reservoirs for infection of hatchery systems. Perhaps we should be more concerned about the converse problem of hatcheries infecting wild fish, particularly in relation to diseases like bacterial kidney disease that develop slowly and hence may not be an obvious target of treatment and prophylaxis within hatcheries.

Predators that are attracted to concentrations of hatchery fish may incidentally take at least some wild juveniles. For example, Wood (1985a, b) found that mergansers are attracted to enhancement releases and may prefer enhanced streams as nesting sites; their young are then prodigious feeders and will impact naturally produced juvenile salmon that are rearing in the streams.

Mixed-stock fishing

Sport and troll fishing effort can respond dramatically to changes in the total abundance of fish available, and it is difficult or impossible to precisely regulate these responses (Argue et al. 1983). Where natural and enhanced stocks are mixed, more hooks in the water will mean higher mortality rates for the natural fish even if the effort is supposedly "directed" at taking hatchery production. The only way that natural fish could be buffered or protected by enhanced fish in such situations would be for the enhanced fish to strike so many hooks as to significantly reduce the average soak time per hook; this effect is unlikely to be significant considering the low catch rates per hook soak time even in very good fishing areas.

Mixed fishing effects offer the simplest explanation for the general trade-off that has occurred in the Pacific Northwest between natural and hatchery production. The scenario is that *(1)* successful hatchery production significantly increased the total abundance of fish at sea; *(2)* sport and troll effort increased in response; and *(3)* higher exploitation rates then caused natural escapements (and, later, recruitments as well) to decline. More recently, declining survival rates in hatchery (and some natural) stocks have exaggerated the effects that step *(3)* alone would eventually have produced (but more slowly—see simulation results in Argue et al. 1983).

POLICY OPTIONS

There are three broad strategic choices for responding to the current state of affairs (Walters 1986, Figure 2.5; Walters and Riddell 1986). The first and most obvious is a holding action where the current enhancement system is maintained and minimal regulations are implemented to prevent further declines in natural stocks; if hatchery survival rates then improve over time (and the ocean environment hypothesis is proved to be correct), it will become necessary to either introduce more

stringent regulations to protect natural stocks or else write off natural production in favor of a hatchery-based system.

The second (and opposite extreme) choice is to declare the hatchery experience a failure except insofar as it has provided mitigation for stocks lost through habitat destruction, and cut back hatchery production while using stringent regulation to rebuild natural stocks to more productive levels. This is a loser's option: short- and probably even long-term catches would be sharply reduced; we would never learn whether the hatchery survival trends would have reversed themselves naturally; and there would be little incentive to seek improvements in hatchery production practices.

The third choice would be some variation on the innovative plan suggested by Ray Hilborn of the University of British Columbia in 1982. He noted that if fixed catch quotas are placed on mixed-stock fishing areas while abundance in these areas is increased due to enhancement, exploitation rates on natural stocks will then decrease; to hold catches at the quota while abundance is increasing, it would be necessary to progressively reduce fishing effort. There is a positive feedback effect built into this policy, as well: after natural stocks start to recover, their increased abundance would drive down the exploitation rate needed to meet the quota still further, which would then promote even more rapid recovery. While this policy might be attractive to commercial fishers (offering the same catch with less effort), it would certainly not be favored by sport fishing interests whose earnings are positively related to total effort rather than catch. The new U.S.-Canada Treaty has essentially adopted a quota plan, though it has loopholes to allow quota revisions in response to increased enhancement production.

Hilborn's concept has many attractive features, especially considering that the basic machinery is now in place for quota regulation of the major mixed-stock fisheries coastwide. Unfortunately, it does not appear that the current enhancement production system can meet his requirement of increasing abundance in the fishing areas; enhancement output has been growing, without producing that desirable effect (in recent years).

At this point we must either admit that holding or natural stock rebuilding policies are the only viable choices, or else begin to look seriously at drastic experiments to find out why the enhancement production system is not working as planned. Suppose, for example, that one major production hatchery is picked from each of the main chinook-coho jurisdictions (Oregon, Washington, British Columbia), and rearing densities (and smolt outputs) for these test systems are cut by 75 percent and kept low for several years. The next year, three additional hatcheries are cut back, and then perhaps (but not necessarily) three more a year later. This "staircase" experimental design (Walters et al. 1986) would test for effects of improved hatchery practice and/or ecosystem responses, while providing controls for changes in oceanographic conditions (many hatcheries continuing present practices) and for "time x treatment interactions" involving the treated hatcheries responding differently to oceanographic changes than the control (standard operating) hatcheries.

The potential cost (in terms of lost production) of an experiment with rearing densities would not be all that great. Suppose that the real problem is oceanographic conditions, and these become favorable a few years into the experiment. The change would be measured by control hatcheries, and the treated hatcheries could be brought back to higher rearing densities with only the few years' partial loss in production. Suppose instead that the real problem is hatchery practices or ecosystem responses (as I would bet); then there would be essentially no net loss in production (and perhaps even a gain) and a substantially cheaper operating regime would have been found that should apply for many other hatcheries. Further, there would be immediate incentive to develop more new (but small-scale) hatcheries. I am not an economist, but it certainly seems obvious to me that the experiment would be a good gamble.

Salmon management in the Pacific Northwest is at a major turning point. We have the choice of trying to continue muddling along with production strategies that are not working well at all, while hoping that natural changes will bail us out, or instead taking some bold steps to try and identify what is going wrong. Those steps should not involve destroying the current enhancement production system, but instead should provide a sequence of localized, controlled changes that can be compared over time to the current system. The key challenge now is to develop institutional arrangements and commitments that will allow us to conduct a sensible and timely experiment.

LITERATURE CITED

Argue, A.W., R. Hilborn, R.M. Peterman, M.J. Staley, and C.J. Walters. 1983. The Strait of Georgia chinook and coho fisheries. Can. J. Fish. Aquat. Sci. Bulletin 211.

Berg, L. and T.G. Northcote. 1985. Changes in territorial, gill-flaring, and feeding behavior in juvenile coho salmon (*Oncorhyncus kisutch*) following short-term pulses of suspended sediment. *Can. J. Fish. Aquat. Sci.* 42: 1410-1417.

Chapman, D.W. 1962. Aggressive behavior in juvenile coho salmon as a cause of emigration. *J. Fish. Res. Board Can.* 19: 1047-1080.

Chilcote, M.W., S.A. Leider, and J.J. Loch. 1986. Differential reproductive success of hatchery and wild summer-run steelhead under natural conditions. *Trans. Amer. Fish. Soc.* 115: 726-735.

Chinook Technical Committee, US-Canada Treaty. 1985. Chinook technical report. Can. Dept. Fisheries and Oceans, Vancouver, B.C. 186 pp. mimeo.

Dill, L.M. 1978. Aggressive distance in juvenile coho salmon (*Oncorhyncus kisutch*). *Can. J. Zool.* 56: 1441-1446.

Fenderson, O.C., W.H. Everhart, and K.M. Muth. 1968. Comparative agonistic and feeding behavior of hatchery-reared and wild salmon in aquaria. *J. Fish. Res. Board Can.* 25: 1-14.

Mason, J.C., and D.W. Chapman. 1965. Significance of early emergence, environmental rearing capacity, and behavioral ecology of juvenile coho salmon in stream channels. *J. Fish. Res. Board Can.* 22: 173-190.

Nickelson, T.E. 1986. Influences of upwelling, ocean temperature, and smolt abundance on marine survival of coho salmon (*Oncorhynchus kisutch*) in the Oregon production area. *Can. J. Fish. Aquat. Sci.* 42: 527-535.

Solazzi, M.F., S.L. Johnson, and T.E. Nickelson. 1983. The effectiveness of stocking hatchery coho presmolts to increase the rearing density of juvenile coho salmon in Oregon coastal streams. Oregon Dept. Fish Wildl., Res. and Dev. Div., Info. Rept. No. 83-1, 14 pp. mimeo.

Walters, C.J. 1986. *Adaptive Management of Renewable Resources*. Macmillan Publishing Co., Inc. N.Y.

Walters, C.J. and B. Riddell. 1986. Multiple objectives in salmon management: the chinook sport fishery in the Strait of Georgia, British Columbia. *Northwest Environmental Journal*, Vol. 2(1): 1-15.

Walters, C.J., T. Webb and J. Collie. 1986. Experimental designs for estimating transient responses to production enhancement measures in Columbia River streams. ESSA Ltd. rept. to Northwest Power Planning Council, Portland, Oregon, 27 p. mimeo.

Wood, C. 1985a. Food searching behavior of the common merganser (*Mergus merganser*) II: choice of foraging location. *Can. J. Zool.* 63: 1271-1279.

Wood, C. 1985b. Aggregative response of common mergansers (*Mergus merganser*): predicting flock size and abundance on Vancouver Island salmon streams. *Can. J. Fish. Aquat. Sci.* 42: 1259-71.

The Culture, Allocation, and Economic Value of Pacific Salmon in the Great Lakes

Howard A. Tanner

Professor of Fisheries
Michigan State University

INTRODUCTION

The successful introduction of three species of Pacific salmon and two species of trout into the fresh waters of the Great Lakes of North America have restored ecological balances and created a recreational fishery currently valued in excess of two billion dollars annually. These changes have been heralded as North America's most outstanding achievement in fisheries management. A complex set of biological, political, economic, and social changes have occurred. Allocation policies have been reversed, and fish stocks are being maintained and expanded by newly built or rebuilt hatchery systems. The large human population in or near the Great Lakes basin has been the source of pollution and contaminant problems, but it also generates a very large recreational demand. The presence of a very large market for recreational fishing opportunities led fisheries managers to shift the allocation of fish stocks away from commercial harvest. This paper is intended to trace the history of events and to describe the complexity of biological, geographical, political, and social relations that have led to the present status of Great Lakes fisheries management.

DESCRIPTION
Physical description and demography

The five Great Lakes and their connecting waterways total almost 95,000 square miles (24.5 million hectares). This watershed discharges at the approximate average rate of 250,000 cubic feet per second into the North Atlantic via the St. Lawrence River. With the exception of Lake Erie, the lakes can be described as deep, cold, and clear with moderate to low rates of biological productivity. Physical descriptions are presented in Table 1.

The connecting rivers have been dredged and waterfalls and rapids bypassed with systems of locks and canals, opening up all five lakes to international shipping. There are also waterway connections to the Mississippi-Missouri River system at Chicago and to the Hudson-Mohawk River system via the Erie Canal. The lakes and waterways are partially frozen during the winter.

The Great Lakes lie close to large populations: 37 million people live in the Great Lakes basin itself and the population of the eight riparian states and the province of Ontario exceeds 84 million. It is estimated that at least 100 million people live within one day's drive (500 miles) of one or more of the Great Lakes.

Political ownership and management authority

The Great Lakes system lies along the U.S./Canadian border, with approximately 69 percent of the lakes' surface area on the U.S. side. Lake Michigan, totally within U.S. boundaries, is the only lake not shared between the two nations. Fisheries management authority lies with the eight U.S. states riparian to the lakes and the single riparian province, Ontario, on the Canadian side. The Great Lakes Fishery Treaty, signed between the two countries in 1955, created the Great Lakes

Table 1. Areas and depths of the Great Lakes.

	Area square miles	Maximum depth (feet)	Average depth (feet)
Superior	31,699	1,335	489
Michigan	22,273	925	279
Huron	22,973	748	194
Erie	9,906	210	62
Ontario	7,336	804	282

Fishery Commission. Sea lamprey control and lake trout restoration were declared to be the responsibility of both nations through federal fishery agencies. Federal fishery research programs contribute substantially to our understanding of the fisheries. The management of certain waters is shared with a number of Native American tribal groups whose treaty rights have been confirmed by recent court decisions.

The states' authority to manage the portions of the Great Lakes lying within their borders is important because it has been the various states and Ontario that have made the decisions to implement major changes in lake management. Federal fishery agencies have frequently opposed and still, in some cases, oppose these decisions. Two political entities, Michigan and Ontario, are in a position to exercise management leadership. Almost 41 percent of the total area of the Great Lakes is in Michigan and 35 percent in Ontario (Figure 1). In addition, over 49 percent of the total area of the upper three lakes—Superior, Michigan, and Huron—is under the jurisdiction of the State of Michigan.

Most of the decisions that brought about today's exciting and economically important recreational fisheries were first initiated by the State of Michigan in 1964, through its Fisheries Division, which decided to undertake management and research programs on the waters of the Great Lakes under its jurisdiction. The keystone of change was the policy decision to make recreational fishery management the primary goal and to relegate commercial fishing to a secondary role. As a result, Michigan's small commercial fishery was reduced in numbers from about seven hundred to two hundred licenses through the procedures of limited entry, retirement of inactive or little used licenses, and buy-out. Today the commercial fisheries of Michigan cannot legally harvest or sell any trout or salmon and must fish with selective gear (chiefly trapnets and pondnets) and in places that do not interfere with sport fishing. Tribal fishers may harvest trout and salmon in certain areas generally removed from the more popular sport fishing areas.

The second important initiative was to recognize that ecological stability could be restored with a number of salmonid species in lieu of or in addition to the native lake trout. Coho and chinook were introduced and steelhead and brown trout populations were expanded to provide fish populations attractive to the recreational angler and to

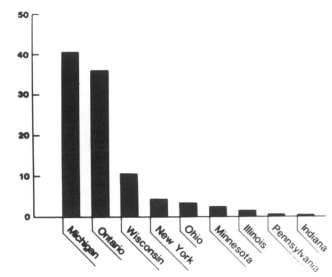

Figure 1. Percentage ownership of the Great Lakes by state and province.

control alewife populations. As the merit of these choices has become apparent the other states and Ontario have developed similar policies and strategies.

HISTORICAL EVENTS

The fish stocks of the Great Lakes remained substantially unharvested until the wave of settlement began after the American Revolution and the War of 1812. By 1850, however, harvest was heavy and many preferred species were showing signs of depletion. Several important species became nearly extinct. In the period between the latter half of the 1800s and the 1950s, the Atlantic salmon, blue pike, and lake sturgeon were eliminated. Lake trout, whitefish, lake herring and most species of chubs were soon overfished in extensive areas of Lake Ontario, Lake Erie, Lake Michigan, and Lake Huron. European settlement also brought clearing of the land for farming and lumbering, causing warming of streams, sedimentation, and periods of reduced flow. Dams were built and many spawning runs were interrupted. Small-scale pollution began.

From the days of the early French explorers the Great Lakes system was recognized as a convenient waterway for travel and transportation. The falls on the Niagara River between Lake Ontario below and the rest of the watershed above were a barrier both to boat traffic and to the upstream movement of fish. In 1828 the Welland Canal around the falls was completed, providing entry into the upper lakes for fishes from Lake Ontario and for anadromous and catadromous species from the North Atlantic. The Erie Canal connecting Lake Erie with the Hudson River and the Trent/Severn Waterways connecting Lake Ontario with Georgian Bay of Lake Huron are also believed to have provided entry for new species into the upper Great Lakes. The principal species to gain access and prominence in fish communities in the upper Great Lakes were the sea lamprey, alewife, and white perch. The American eel also achieved entry but has never been present in significant numbers.

These events began the process of weakening the ecological stability of all or most of the Great Lakes fish stocks. Even prior to any exploitation, the ecological relationships may not have been particularly stable since these lakes had emerged as the glaciers retreated 10,000-15,000 years ago and the time for diversity and niche occupancy to develop was limited. Probably no more than 150 species of fish were present and many of these had limited distribution and/or were numerically unimportant.

Another form of ecological disruption occurred as people began to successfully transport a number of fish species. Between 1870 and 1920 the carp, smelt, rainbow trout, and brown trout were successfully introduced into the upper Great Lakes systems. Both the carp and the smelt became very abundant and, in the opinion of most fishery scientists, did so to the detriment of native fish species. In the Great Lakes the predacious food habits of the smelt, especially in its cyclical periods of extreme abundance, have severely depressed the survival of fish eggs and larvae of associated species.

The stability of predator-prey relationships was further damaged and the assured levels of recruitment and year class strength further assailed by events of the 1940s and 1950s. High levels of harvest were caused by the increased levels of gear efficiency brought about first by nylon twine and then monofilament; increased numbers of workers following the end of World War II; and increased demand and higher prices. Collapse of lake trout and whitefish stocks followed. The population explosion of the parasitic sea lamprey occurred during the same period and no doubt hastened the collapse of these fisheries. However, its role has probably been exaggerated in order to obscure the responsibility of those who shared in excessive harvests; it was a convenient scapegoat.

By the early 1950s the lake trout, the largest and most important predator, was gone from Lakes Michigan and Huron and its abundance was greatly reduced in Lake Superior. Many commercial fishing enterprises ended while others shifted to the smaller species of chub, herring and, in certain locales, yellow perch. These species also were rapidly depleted. Several species of chub became extinct and the herring disappeared except from Lake Superior and northern Lake Huron.

During these same two decades increased industrial activity and the beginning of the chemical age added higher levels of pollution, particularly along the southern shores of the Great Lakes. The advent of phosphates as an ingredient in detergents and the accompanying loss of oxygen (from enrichment) in portions of Lake Erie was also a phenomenon of this period.

One last but very important ecological change was a part of the Great Lakes historical tragedy. This was the population explosion of the alewife in Lakes Michigan and Huron. This small invader from salt water had made its way into the upper Great Lakes following the access provided by the various canal systems. The alewife, following decades of being present in insignificant numbers, produced a population explosion in the early 1960s. The disappearance of the predacious lake trout and the decline of chub and herring populations, the alewife's probable competitors for zooplankton, may well have aided in its proliferation. The lakes were increasingly polluted by industry and sewage and contaminated by pesticides. The commercial fisheries were mostly bankrupt and, perhaps equally important, they were politically impotent. The time was ripe for change.

In 1955 Canada and the United States had signed the Great Lakes Fishery Treaty, which created the Great Lakes Fishery Commission. This Commission, without management authority, nevertheless has major accomplishments to its credit. Very early it selected sea lamprey control and lake trout restoration as its highest priorities. Years of research subsequently led to the discovery of an effective lampricide. The application of this material to lamprey spawning streams on both sides of the border has produced reasonably effective control and has reduced lamprey numbers by at least 90 percent.

With several hundred prime spawning stream systems to be treated, this program is an expensive management cost that in 1985 totalled approximately six million dollars of state and federal funds on the U.S. side and probably a comparable expenditure on the Canadian side. Effective sea lamprey control began on the Lake Superior drainage in about 1959, proceeding in subsequent years to Lake Michigan, then to Lake Huron, followed by Lake Ontario. The sea lamprey populations of Lake Erie were never large and spawning streams limited, and control on this lake has only been initiated recently. The research leading to a system of effective sea lamprey control was one of the most significant achievements in the professional fisheries field and continued control remains essential to any salmonid management program in the Great Lakes.

The lake trout restoration program, the second priority set by the Great Lakes Fisheries Commission, was perceived as essential to restoration of predator-prey balance and as a primary element in restoring a viable commercial fishery. Today, 30 years later, lake trout continue to be stocked in very large numbers by both countries; approximately 140 million lake trout fingerlings have been stocked in the last 25 years. However, the goal of a self-sustaining population throughout original lake trout habitat remains largely unfulfilled. Lake trout stocked as fingerlings have a satisfactory survival rate but only in Lake Superior have these stocked trout contributed to natural reproduction. Elsewhere naturally spawned lake trout have been detected in very limited numbers and in limited areas. The reasons for the lack of success are debatable and not understood. Stocked lake trout make up a significant percentage of the salmonid sport catch and are taken by tribal commercial fishers in certain waters of some states. Commercial fisheries remain dominant in many Canadian waters and stocked lake trout make up significant portions of catch value.

The goals of Great Lakes management have, however, changed since the inception of the lake trout restoration program. At first, it was intended that the ecological balance of the lakes should be restored by the use of only native species—namely, lake trout—and that the fisheries would be restored primarily for commercial harvest. Since that time, allocation of the harvest to recreational fishing and an ecological balance using a group of exotic salmonids more attractive to recreational fishers have replaced the original goals. This shift, initiated by Michigan and subsequently implemented by the other states, has produced a set of political conflicts. The existing emphasis on commercial fishery allocations was supported by the Great Lakes Fishery Commission and by federal agencies in both the U.S. and Canada, while the shift to sport fishing objectives was led by the fishery agencies of the states. On the U.S. side, these conflicts are substantially past. The commercial fisheries, weakened politically by years of near bankruptcy, have yielded to the policy decisions that relegate them to a limited and secondary role. Commercial harvest in much of U.S. waters may occur only in a manner that does not adversely impact on recreational harvest. Tribal commercial fisheries remain an exception to this but have, at least to date, been restricted by agreement to areas where their impact on sport fishing is limited.

Whether lake trout can be restored to self-sustaining populations remains questionable; the program's success is more assured in Lake Superior and only a possibility in the other lakes. With the salmonid population shifted to other species and with recreational fishing a primary goal, the appropriateness of lake trout stocking and restoration is questionable. Ecologic balances have been achieved and can be sustained with other species that grow and mature faster and are more highly regarded by sport anglers. The lake trout, less efficient in food conversion and slow to mature sexually, remains a very difficult species to manage.

IMPETUS FOR CHANGE IN MICHIGAN

The sweeping changes in fisheries management that occurred in Michigan in 1964-66 were a part of one of those rare periods of history when an old way of doing things is abruptly and thoroughly changed. In response to needs that had accumulated as a result of a very prosperous period of industrial expansion and human population growth, the Michigan State constitution was rewritten by a convention in the early 1960s. The well-being of the state's natural resources was declared to be of paramount importance. A Department of Natural Resources (DNR) was authorized, bringing together all elements of public resources, land and water management, environmental regulation, and public recreational resources. A new governor campaigned on a pledge to modernize the yet-to-be-created DNR. A new professionally trained director hired new division chiefs, and professionally trained fishery biologists for the first time gained leadership positions within the Fishery Division.

The new Great Lakes fishery management initiative was implemented by a series of sweeping changes. The oldest and least efficient hatcheries were closed. The program of stocking trout of catchable size for a quick harvest was terminated. Hatchery space and about one million dollars were thus made available.

No fishing license had ever been required for recreational fishing in Michigan waters of the Great Lakes, but in 1967 legislation was passed requiring a fishing license and later a trout and salmon stamp. Trout and salmon in the Great Lakes were declared to be game species not subject to commercial fishing. Commercial licenses not qualifying as full-time or principal occupation were terminated. Large mesh gillnets were outlawed and commercial fishers were permitted to fish only with selective gear (principally pond and trap nets for whitefish) or small mesh gillnets for chub in over 40 fathoms (about 240 feet) of water.

These changes were accompanied by an intense and sustained public relations campaign that emphasized three elements. One, the dead and dying alewives that each year littered beaches and plugged water intakes were not to be viewed as a problem but as a forage base upon which to build a recreational salmon and trout fishery. Two, the then current policy goal of the Great Lakes Fisheries Commission and the U.S. Fish and Wildlife Service (then represented in the Great Lakes by the Bureau of Commercial Fisheries) to reestablish lost commercial fisheries was criticized as a misguided effort to turn back the clock. Michigan fishery leaders declared that their goal was to create a recreational fishery to serve the demands of millions of nearby recreational anglers. Three, the best way to do this was with species of trout and salmon that were highly regarded by sport anglers, could be expected to fit within the ecology of Great Lakes waters, and would eat alewives.

INTRODUCED SPECIES OF TROUT AND SALMON

Contrary to the belief held by the public and even by some biologists, most species of Pacific salmon do not need to spend a part of their life cycle in salt water. Coho and chinook salmon were the major species introduced into the Great Lakes in the 1960s, but the earlier presence of three other exotic salmonid species needs to be explained.

Steelhead (rainbow trout) had been introduced into Michigan and other Great Lakes states in the 1870s and, while not rare, neither did they support any sustainable harvest. Brown trout, another exotic salmonid species, had also been introduced into Michigan in the 1870s. This species has been consistently stocked in many streams and lakes but was present only incidentally in Great Lakes waters.

The introduction of pink salmon was not an integral part of the story of how the Great Lakes recreational fishery was established, but it is a story worthy of digression. In the mid-1950s pink salmon were reared in a Canadian hatchery within the Lake Superior watershed for intended release in one or more streams tributary to Hudson Bay. An unknown but very small number either escaped or were released either into Lake Superior or a tributary. In 1959 a few fish spawning in a Minnesota Lake Superior tributary were collected and identified as pink salmon. Thereafter, principally in odd-numbered years, spawning pink salmon were observed spreading through the Great Lakes. The increase and spread of pink salmon is, therefore, entirely the result of natural reproduction.

In recent years pink salmon have been observed in tributaries to all five of the Great Lakes. They are now spawning every year. A small commercial fishery has developed in the Canadian waters of Lake Huron (300,000 pound quota in 1986). Pink salmon in the Great Lakes are small. Recent studies have shown that they feed extensively on small fishes as well as on plankton. In 1985 pink salmon were especially abundant in Lake Huron and were a nuisance to sport anglers.

The release of pink salmon into the Great Lakes was unplanned and unauthorized. Their ultimate position in the Great Lakes fisheries ecology is as yet unclear. The tentative judgment is that they will prove to be a largely unwanted competitor and small fish predator.

The first intentional introduction of the 1960s occurred in the fall of 1964, when kokanee eggs were obtained from Colorado for stocking in inland trout lakes. This was not precisely a precursor to Great Lakes stocking of salmon, but it did serve to introduce the public to the concept. Later in 1964 it was learned that surplus coho eggs were available in Oregon and Washington. The coho had already been selected as the first species to be introduced and the first one million eggs were shipped to Michigan from Oregon in late 1964. In subsequent years coho eggs were also obtained from Alaska and Washington. In 1968 the first chinook eggs were received from the state of Washington (Toutle River Hatchery). These early contributions were supported by additional eggs until Michigan was able to become self-sufficient.

Certain conditions were attached to the first shipment of eggs, on the advice of experienced salmon biologists. These stipulated that the coho be raised to smolt size, 5 to 6 inches, on the Oregon Moist Pellet Diet. It was further stipulated that they be released in streams where they had been imprinted and that no less than 250,000 were to be released at each planting site.

From the one million eggs, 850,000 coho were raised to smolt size and released at three sites in the spring of 1966. The first release was at the rearing station on the Platte River, Benzie County, near the town of Honor, Michigan. The date was 2 April 1966. That same day releases were made in creeks tributary to Bear Creek, itself a tributary to the Big Manistee, Manistee County. Both the Big Manistee and the Platte are tributary to Lake Michigan. These releases totalled approximately 670,000 smolts. Later that spring approximately 180,000 smolts were released near the mouth of the Big Huron, a tributary to Lake Superior. This later planting was a concession made to certain legislators to assure funding for the initial salmon rearing.

Public interest in the success of the initial coho stockings was maintained through a sustained publicity effort. In the fall of 1966 an estimated 10,000 precocious males were caught by steelhead fishers in several Lake Michigan tributaries. The size of these coho surprised everyone. The average weight was 3 to 4 pounds with a number of fish recorded in excess of 7 pounds. "Coho fever" began. In the spring of 1967 more coho were released in Lake Michigan and Lake Superior. (No releases were made in Lake Huron at this time because sea lamprey control was not yet in place. With hindsight, this precaution was probably not necessary.)

Fishers waited impatiently until the first coho were caught in Lake Michigan in August of 1967. Numerous coho exceeded 20 pounds! The enthusiasm of the inexperienced Michigan salmon fishers erupted. Facilities of port communities were swamped as thousands of fishers descended on the port cities of Ludington, Manistee, and Frankfort. The excitement is hard to describe, but what was to become known as "salmon fever" was rated in Michigan as one of the most newsworthy events of 1967.

The ill-prepared, inexperienced anglers caught a relatively small number of coho in the open water of Lake Michigan. Coho entered the streams, principally the Platte and Manistee, by the tens of thousands. Eggs were successfully collected and hatched by the Michigan Fisheries Division and Pacific salmon in the Great Lakes were a reality. The final returns—coho caught by fishers or collected at temporary weirs—indicated an astounding 36 percent survival of the original plantings. This high rate of survival undoubtedly reflected the almost complete absence of predacious fish, limited angler harvest, and the extreme abundance of food, principally the alewife.

The acceptance of salmon introductions into the Great Lakes benefited from a public misconception that developed in 1967. The problems of dying alewives littering the beaches, plugging water intakes and, with their stench, causing the evacuation of hundreds of miles of tourist beaches, had been an increasingly severe problem through the summers of the mid-1960s. Every community with tourist beaches was suffering important economic losses and all were screaming for relief. During the summer of 1967 the alewife populations of Lake Michigan collapsed. Billions of dead alewives littered 300 miles of Lake Michigan beaches and dead floating alewives in patches as much as eight miles long were documented by pilots. That fall the first mature coho were caught by the delighted fishing public, and wonder of wonders, the next summer there was no problem with dead alewives. The public was instantly convinced—the coho had proved to be the solution to the alewife problem just as the DNR had said they would be! Fisheries biologists understood that the coho surviving from a planting of 670,000 in a lake of 23,000,000 acres could not possibly have made a dent in alewife numbers. They explained that an exotic species typically expands until carrying capacity is exceeded, at which point a population collapse follows, so that the 1967 die-off represented the predictable collapse of the alewife. But the public was convinced that the coho were responsible. From that

time, public support of salmon stocking, later expanded with steelhead and brown trout, has been assured.

In 1968 and 1969 some record-breaking coho, probably 4-year-old fish, were recorded. A world record of over 39 pounds was taken, with several more in excess of 35 pounds. Unfortunately for Great Lakes history these were not taken on hook and line but were a part of the surplus harvest at harvest weirs operated by the DNR. After the period of initial excitement had passed, the hard work of expanding the fishery to its full potential began.

The Fisheries Division of what by then had become the Michigan Department of Natural Resources was not funded, staffed, or equipped to research and manage the Michigan waters of the Great Lakes, the task that it had undertaken. Prior to 1964 the Department had managed only the inland waters of Michigan, equal to only 3 percent of Michigan's Great Lakes waters.

Funds were increased, appropriations for new hatcheries fought for and secured. Hatcheries, research stations, and vessels were built and staffed. Laws were strengthened. Problems of toxicants in fish, snagging violations, lawsuits with the waning commercial interests, the turmoil created over federal court cases recognizing special status for tribal members to fish—all these and more were dealt with and expansion was rapid and sustained.

The chinook, first obtained in 1968 from the State of Washington, have become the most abundant and popular species. Steelhead have rebounded from low levels of the 1950s. Natural reproduction of steelhead is extensive in scores of Great Lakes streams, and is expanded further by the stocking programs of several states. New strains of steelhead have been introduced. The most popular of these is the Skamania, first introduced by Indiana and then by Michigan. Brown trout are stocked extensively by Wisconsin, Michigan, and New York. Lake trout continue to be reared by the two federal governments and shared with the states and Ontario. Atlantic salmon from Canada, Sweden, and New York, both sea-run and land-locked varieties, have been a part of the stocking program of several Great Lakes states for a number of years, but success has been minimal. The development of an adequate brood stock has not occurred in spite of many years of diligent effort. Michigan and perhaps other states continue to attempt large-scale rearing and release of this very desirable species.

Atlantic salmon are occasionally taken by fishers; however, numbers remain small, insignificant in the total catch.

The patterns of hatchery development and the stocking of the Great Lakes by Michigan have been followed by all the other states and Ontario. Michigan has shared its egg production facilities and supplies all of the chinook and coho and some of the steelhead eggs for the other states. Lake Ontario has become an important recreational fishery for salmon and lake trout. The only chinook in excess of 40 pounds taken in recent years have come from Lake Ontario.

CURRENT STATUS OF THE FISHERY
Salmon/trout stocks

Hatchery production capable of stocking all suitable portions of the Great Lakes with the optimum number of salmonids has been nearly achieved. The limiting factor to the Great Lakes trout and salmon fishery is the forage base available. All estimates indicate that we are at or very near carrying capacity. It is possible that we have exceeded optimum stocking rates in the instance of Lake Michigan, where a maximum number of salmonids —about eighteen million—were stocked in 1984. This number has been reduced by agreement to about fourteen million in 1986. Fisheries managers have the important advantage that, since the majority of salmonids are of hatchery origin, stocking rates can be adjusted annually to closely match the best estimate of carrying capacity. The social/political aspect of these adjustments must be recognized. Every fishing community will lobby for the largest possible stocking rate in their stream or in nearby waters. The sum of each community's goals will predictably exceed carrying capacity. Unrealistic demands must be effectively resisted by the managing agencies.

Natural reproduction of chinook and steelhead makes an important contribution to total annual production of juveniles. Since many scores of streams are involved, it is difficult to estimate the numbers of recruits. However, Michigan has determined that chinook reproduce successfully in most stream areas of suitable quality accessible to anadromous fish. This includes over a hundred streams of trout water quality and some streams of lesser quality. It is estimated that 15 to 20 percent of the chinook present in the upper three lakes are from natural reproduction.

The steelhead has spawned successfully in many Great Lakes streams for perhaps a century or more. Total numbers are not estimated but are a very significant part of the steelhead stocks. Natural reproduction by coho is known to occur in many streams but appears to be very limited. Lake trout reproduction is well established in Lake Superior but insignificant elsewhere. Brown trout reproduction by spawners from the Great Lakes is assumed but is masked by the spawning of stream resident brown trout populations.

Forage base

The period when alewife populations dominated the community of plankton-feeding fish ended perhaps a decade ago. Alewife abundance has declined further and in Lake Michigan in 1985 was estimated to be about 12 to 15 percent of its former level. Alewife prominence in salmonid diets has declined sharply. Several of the native species that had declined to seriously low numbers during the period of high alewife populations have begun to recover. The bloater chub has nearly doubled its population annually for the last several years. The yellow perch is at record high levels in Lakes Michigan and Huron. In the last 3 years yellow perch have been recorded with increasing frequency in the salmon diet. The emerald shiner, once prominent in the Great Lakes, has returned to significant populations following years of being nearly absent. Several other species of cyprinids are now appearing in trout and salmon stomachs. The smelt, always important as a forage fish, appears to be at higher levels of abundance but populations fluctuate dramatically from year to year. Two species of sculpin occur as diet items, principally in lake trout.

Sport fishing on Lake Michigan

The Great Lakes sport fishing harvest of salmon and trout is perhaps larger than generally recognized by people outside the region. It may be also unique in that there is a consistent surplus of fish. This necessitates a system of harvest weirs to remove stream migrants where they appear in numbers in excess of spawn-taking requirements and stream sport fishing opportunity.

In reviewing comparative creel census reports from the various management entities, I have found it impossible to present a composite picture of the sport harvest for the whole Great Lakes. I have therefore opted to present data from Lake Michigan alone. The Lake Michigan fishery is without doubt the largest, both in terms of angling effort and of fish caught. However, that is not to say that the trout and salmon fisheries in the other lakes are not large and important.

In order to compile estimates of an annual sport fishing harvest (Table 2), I have used data from three different years, 1983, 1984, and 1985 (personal communications from G. Rakoczy, Fisheries Division, Michigan DNR, who supplied both Michigan and Wisconsin data; D. Brazo, Division of Fish and Wldlife, Indiana DNR; and R Hess, Fish and Wildlife Resources Division, Illinois Department of Conservation). The result should be judged not as the catch from any one year but rather as a reasonable representation of a typical mid-1980s year. The data presented in this table include the trout and salmon caught in open water, from piers, from the surf, and the stream catch for the states of Wisconsin, Illinois, and Indiana, while the data from Michigan represent only the open water catch. Because there are so many spawning streams and so many miles of beach and piers in Michigan, it has not been feasible to attempt a creel census except for the open water (boat) fishery. For this reason the totals represented in Table 2 are clearly an underestimate of the actual sport harvest. It should be noted that only with the Michigan data were confidence limits available.

Chinook salmon are the most abundant and are dominant with an annual sport harvest of 934,000, making approximately 48 percent of the catch. They are followed by lake trout at 20 percent, coho at 18 percent, brown trout at 7 percent, and steelhead (rainbow) at approximately 6 percent. The total estimated sport harvest of trout and salmon from Lake Michigan for one "typical mid-1980 year" was 1,917,000 fish.

Table 2. Estimated annual sport fishing catch of salmonids, Lake Michigan.

	Michigan[a]	Wisconsin	Illinois	Indiana	Total
Chinook	511,400 (50,100)	328,000	65,000	29,700	934,100
Lake trout	142,200 (17,800)	220,000	13,000	3,600	378,800
Coho	112,000 (15,100)	59,600	165,000	29,900	366,500
Steelhead	46,800 (9,100)	23,200	8,500	33,300	111,800
Brown trout	48,900 (7,300)	72,200	3,500	1,900	126,500
Totals	861,300	703,000	255,000	98,400	1,917,700

Notes: Michigan and Illinois catch data are from 1985; Indiana data from 1984; Wisconsin data from 1983. All data rounded to the nearest hundred.

[a] Confidence limits of two standard errors shown in parenthesis.

For the sake of completeness it is desirable to add some understanding of the magnitude of the most important missing data, the harvest by anglers fishing the shores, piers, and streams in Michigan. Estimates of the harvest by sport anglers from five Michigan streams have been made (personal communication, G. Rakoczy, Fisheries Division, Michigan DNR). These five include some of the largest, most productive streams, but also two relatively small streams—the Betsie and the Platte. The data on the five streams presented in Table 3 are intended to show that there is significant catch unreported in Table 2. However, there is no acceptable way to obtain an estimate of the total angler harvest from Michigan streams by extrapolation. When the contents of Tables 2 and 3 are combined, the sport catch of salmonids from Lake Michigan and its tributaries is shown to exceed two million fish and the catch of chinook alone exceeds one million, excluding the catch from many Michigan streams where no angler census was attempted.

Table 3. Anglers' harvest from five Michigan streams, 1984.

	St. Joseph River	Muskegon River	Big Manistee River	Betsie River	Platte River	Totals
Chinook	4,142	20,631	61,492	10,687	472	97,424
Coho	212	281	5,478	3,079	1,176	10,226
Steelhead	2,513	8,935	20,162	3,117	863	35,590
Brown trout	1,423	418	813	790	2	3,446
Totals	8,390	30,265	87,945	17,673	2,513	146,686

The average size of salmon caught in the Great Lakes is probably smaller than salmon caught from the Pacific Ocean and its tributaries; it has declined as stocking rates have increased. Average weights are available for Lake Michigan from the 1984 catch in Wisconsin and from the 1985 catch in Michigan (Michigan figures do not include the fall stream fishery) and are presented in Table 4.

The average weight of chinook caught by anglers is approximately 12 pounds; it is clear that sport anglers harvest over 12 million pounds of chinook from Lake Michigan and its tributaries each year. Recall again that this is an underestimate because of the lack of creel census data from many Michigan streams.

In order to complete the perspective on salmon harvested annually from Lake Michigan and its tributaries, three other catch elements need to be mentioned. These are the incidental mortality of salmon in legal commercial gear, the harvest by tribal fishers, and the catch of chinook and coho at harvest weirs. It is unforgivable that Indiana, with approximately 40 miles of shoreline, enjoys the benefits of a very large sport fishery (based largely on fish stocked by other states) and yet continues to license nineteen commercial gillnet fishers. The loss of salmon and trout to this gillnet fishery (largely for yellow perch) in Indiana waters exceeds 75,000 chinook each year (D. Brazo, Indiana DNR, personal communication). These trout and salmon cannot be legally sold.

The tribal commercial fishery is mostly a large mesh gillnet fishery with some impoundment nets. Whitefish are their principal target fish. They can legally sell trout and salmon. More complete catch data on the tribal fishery will become available in future years.

Catch at harvest weirs is the last element of the total annual harvest from Lake Michigan. The Michigan DNR has permanent harvest weirs on the Little Manistee and Platte rivers. Here eggs are collected annually for all management agencies requesting coho and chinook eggs. The requests may exceed forty million, not counting the spring take of steelhead eggs. The weirs are operated by a private contractor under the supervision of the Michigan DNR and the carcasses and surplus eggs are processed and marketed by this contractor under a competitive bidding arrangement. The contract also requires that the contractor erect and operate temporary weirs on several other Lake

Michigan tributaries. In all weir operations, surplus salmon are harvested and steelhead and brown trout passed upstream. On some streams and in accordance with instructions from the Michigan Fisheries Division, predetermined numbers of salmon are allowed to proceed upstream. The numbers of salmon harvested at Lake Michigan weirs in 1985 are as shown in Table 5.

The preceding data from Lake Michigan permit the reader to judge the magnitude of the sport catch and elements of harvest of trout and salmon from that one lake during a typical year. Lake Michigan does have the greatest number of anglers and number of fish caught of any of the Great Lakes. However, the sport fishery for trout and salmon in the other lakes and other sport fisheries for walleye, yellow perch, and several other species are large and growing because the available fish stocks have in most Great Lakes waters been allocated to sport fishing.

Estimates of angler effort are limited and difficult to total for the four states riparian to Lake Michigan. However, from partial creel censuses, records of charter boat catch, and licenses sold, it is estimated that at least forty million angler hours are expended in pursuit of trout and salmon on Lake Michigan and its tributaries.

The management rules for sport anglers on Lake Michigan recognize that a surplus exists and that it is desirable to increase the sport catch; as long as surplus numbers appear in the spawning streams this will remain true. Sport fishing regulations on Lake Michigan are similar in the four states represented: Michigan, Wisconsin, Illinois, and Indiana. The season is open year round. However, the severe weather of winter limits fishing activity, generally to the period of April through November. The only exception to a year-round season is on lake trout where efforts continue towards a self-sustaining population. Lake trout may be retained by sport anglers only between May 15 and August 15.

The daily creel limit is five trout and salmon in combination, only one or sometimes two of which may be a lake trout. There is no season or annual limit. Anglers are limited to two rods per person with no more than two baits/lures on each rod. The size limit is 10 inches, but fish this small are rarely encountered except perhaps by stream anglers.

Regrettably, the large number of salmon returning to the streams has created a serious problem with illegal foul hooking or snagging. Snagging is legal at five locations in Michigan. Three of these are at dams where there are no functioning fish ladders and large numbers of salmon accumulate.

Nearly all sport fishing on the Great Lakes is with artificial lures. Temperature, depth, structure, and bait schools are all important in determining the location of salmonids. The fish are often at depths from 65 to 125 feet in midsummer. To present lures at the proper depth, a downrigger, either electric or manual, is employed. A downrigger, invented on the Great Lakes, is a small winch with calibration to allow the trolling of a 8- to 12-pound "cannonball," to which lines and lures are attached by means of a variety of releases. A strike by the fish releases the line and lure from the "cannon ball" and the fish and angler are connected without any weights on the line. Additional equipment typically found on Great Lakes fishing boats includes fish finders, outriggers, two-way radios, trolling speed indicators, and constant read-out thermometers. Occasionally there may be auto pilots, radar, and Loran-C equipment.

Harbor Facilities

Michigan began to expand its boating facilities more than 20 years ago. One initial goal was to supplement natural harbors with a system of harbor refuges no more than 15 miles apart throughout the 3,200 miles of Michigan's shoreline. This system is essentially complete. Additional facilities continue to be built or expanded with a financial support system that typically incorporates federal, state, and local government funding. Funds are largely from taxes on marine gasoline sales. Boat slips and launching facilities continue to be in short

supply. Whether built by public or private sources, boat slips always have a waiting list, often up to 3 years. The total number of boats (over 14 feet in length) registered in Michigan is over 700,000, more than any other state, or about 8 to 9 percent of the boats registered in the whole country.

Numbers of charter boat licenses have expanded rapidly. In 1986 Michigan had 930; Wisconsin over 500; Illinois 225; and Indiana 82; for a total of at least 1,737 charter boat licenses.

The economic impacts on those communities serving as principal ports for the recreational fishing demand is indeed dramatic. In Michigan along the Lake Michigan shore there are about twenty principal ports. Each of these have followed similar response patterns. The Michigan geology is such that most tributaries to Lake Michigan have what is termed a "flooded river mouth" represented by a lake of varying sizes just inland from the final exit channel to Lake Michigan. Most of these were dredged and channeled into harbors many years ago. Small boat facilities are continually being expanded in all of these to meet the ever-increasing demands of recreational fishers and other recreational boaters. While this may be a redistribution of wealth rather than new wealth, the economic growth and vigor of these communities must be seen to be appreciated. Estimates of the total economic impact of the rebuilt and introduced stocks of salmon and trout upon the sport fisheries of all five of the Great Lakes vary. The total economic activity is most often stated at between one and a half to two billion dollars annually (Talhelm 1981). The highest figure I have seen quoted is four billion dollars. Whatever the accurate figure for economic activity generated, it is obviously very large and still growing very rapidly.

Table 4. *Average weights of sport-caught salmonids from Lake Michigan.*

	Wisconsin	Michigan
	(pounds)	
Chinook	13.5	11.3
Coho	4.3	4.7
Lake trout	7.6	6.0
Rainbow	6.5	5.8
Brown trout	5.8	5.6

Note: Wisconsin data are from 1984; Michigan data are from 1985.

Table 5. *Salmon and trout harvested or passed upstream at weirs on two Michigan rivers, fall of 1985.*

	Platt River	Little Manistee River	Totals
Chinook	3,000	34,000	37,000
Coho	74,000	15,300	89,300
Steelhead[a]	1,200	6,400	7,600
Brown trout[a]	114	177	291

[a] All steelhead and brown trout passed upstream.

THE FUTURE

The achievements of Great Lakes fisheries management over the last 20 years provide a basis for optimistic predictions. The capability to stock the lakes at carrying capacity is in place or nearly so. The species of trout and salmon that support our sport fishery are essentially in place. Since they are principally of hatchery origin, numbers can be adjusted annually to the forage base. Fishing pressure will no doubt continue to increase and eventually additional catch restrictions will become necessary.

As I conclude this generally optimistic paper, it is worth mentioning that a controversy over management of Lakes Michigan and Huron is not far off. Fishers are increasingly aware that the salmon in their catches are smaller each year. These Great Lakes anglers are most apt to attribute this decline in size to a decrease in alewife and other forage species. Michigan biologists have adjusted stocking rates downward and, in examining the length and weight of chinook returning to the weirs, have noted no decline in average size of the returning mature chinook during the last 4 years. What appears likely is that the ever-increasing exploitation rate by sport anglers is resulting in a decline in the average age at harvest. Certainly, theory would lead us to expect this.

How this problem will be met is a matter of concern and probable dispute in the years ahead. Certainly, our fishery management agencies can restrict seasons and creel limits, impose size limits, and reduce annual harvests. If and how these measures are employed, how they are balanced with the stocking rates and available forage are predictably some of the very important and challenging tasks for fisheries managers in the years ahead.

The states of Michigan, Wisconsin, and New York are experimenting with a novel solution to the problem by producing sterile chinook. Sterility is being achieved in Michigan through the use of heat shock on newly fertilized eggs to produce triploidy. The first triploidy smolts were marked, tagged, and released into Lake Michigan and Lake Huron in April of 1986. The envisioned goal is to extend the life cycle of some chinook, thereby producing a few larger fish to add to the excitement of any Great Lakes fishing experience.

The issue of catch size clearly adds pressure to eliminate any remaining commercial harvest of salmonids. In addition, the political lobbying of affected communities, each seeking increased stocking in their vicinity, will make management decisions more politically complicated.

The demand for boating facilities continues to expand faster than they can be built. Additional and improved facilities are being built as rapidly as possible. The communities serving Great Lakes anglers are competing to provide the most satisfactory fishing experience. Many communities have also recognized that the nonfishing members of visting groups must also have an enjoyable experience and are enhancing their shopping, beaches, local attractions, and other recreational facilities.

Some problems remain unsolved. Various tribal fishing issues remain in the courts. However, the majority of the areas in dispute lie well to the north of the more important sport fishing ports.

The gene pool of chinook in Michigan is particularly limited. The original source was the Toutle River Hatchery in Washington. Michigan has taken eggs from the progeny of these fish and shared them with the other states and Ontario. Several other strains of chinook appear to have different and very desirable traits. Currently some stocks of spring run chinook are being examined for introduction. The cooperation of west coast fisheries managers has been very helpful and hopefully will continue. The presence of serious fish disease problems is a significant barrier to additional introductions. A number of troublesome diseases present in many Pacific stocks are not currently present in Great Lakes trout and salmon. Obviously, rigorous inspections and certification are in order.

The problem of contaminants, DDT, PCBs, dieldrin, toxophene, etc. have been a serious problem for both the sport and commercial fisheries of the Great Lakes. However, the decline in the concentration of all of these various materials is very encouraging. The restrictive rules on the use/discharge of these compounds have been effective in reducing concentrations throughout the Great Lakes system. Some, like DDT, are now close to zero; all have declined substantially and very few restrictions remain on the sale of fish from the Great Lakes. Large lake trout with unacceptable levels of PCBs are the only significant exception. Much of the remaining input of persistent chemicals has been shown to be airborne and often transported from great distances. These sources are difficult to eliminate but will decline further as the countries of North America begin effective measures to reduce airborne contamination.

Creation of a sport hatchery for trout and salmon in the Great Lakes is a development of great significance. This fishery is founded on several basic factors. The physical and biological traits of the Great Lakes are well suited for trout and salmon, with the exception of most of Lake Erie. The management agencies, one after another, have opted to allocate trout and salmon to a sport fishery and to greatly restrict or eliminate any commercial fishery that negatively affects this allocation. The parasitic sea lamprey has been brought under control. The salmon and trout have a large and diverse forage base. Natural reproduction contributes to a limited degree while hatchery produced trout and salmon provide the majority of the catch. This permits reasonably precise adjustments to lake carrying capacity. Supporting facilities for fishers and their boating needs continue to expand towards fulfillment of as yet unmet demands.

From an unbalanced, overfished, parasitically plagued disaster, Great Lakes fish stocks have been rebuilt and reallocated to a sport fishery that continues to grow and serve more people. The sport fishery for trout and salmon has become a major economic asset both to local areas and to the region as a whole. Economists attribute from two to three billion dollars in economic activity to Great Lakes sport fishery. It is reasonable to expect the value of this fishery to continue to expand.

LITERATURE CITED

Talhelm, D.R., et al. 1979. Current estimates of Great Lakes fisheries values: 1979 status report. Great Lakes Fishery Commission, Ann Arbor.

Conservation and Allocation Decisions in Fishery Management

Courtland L. Smith

Professor of Anthropology
Oregon State University

WHAT IS FISHERY MANAGEMENT?

Fishery management is the dual process of conservation and allocation. Conservation maintains resource productivity for future generations. Allocation controls participation in a fishery.

Conservation[1] asks: *How much?* It involves a determination of the amount of a fish stock available for use while also keeping the population healthy and viable into the future. The maintenance of fish populations for present and future generations is a broadly held social goal. A holistic view is necessary for conservation because the viability of fish stocks is dependent on other habitat uses. These include water supply, waste water disposal, watercourse modifications and obstructions, and competing water use activities. The social priorities of these activities extend beyond the narrower concerns of fishing. Thus, conservation organizations with the most encompassing authority have the greatest success.

Allocation involves a determination of *who* gets the opportunity to participate in the fishery. It is the who, when, where, and how of fishery management. Allocation can be an active process or it can occur, de facto, by inaction. Allocation distributes the opportunity to catch fish among users.

Some argue that conservation, because it is a "how much" question, can be determined by scientists, particularly biologists. On the other hand, it is argued, allocation, because it is a "who" question, is more social and political. This is a false dichotomy. Conservation is not a political issue because there is strong public backing for maintaining selected fish populations. Allocation becomes politicized because there is less social and political agreement on appropriate goals than there is for conservation.

WHY DISTINGUISH CONSERVATION FROM ALLOCATION?

The "tragedy of the commons"[2] refers to the problem of heavily exploited fisheries with open access. Increasing numbers of people seeking an allocation from the fish stock threaten the stock's future productivity. Economic and social incentives to catch more fish cause overexploitation. In these situations, separating conservation from allocation is necessary to protect the resource. The purpose of separating conservation and allocation is to give priority to the maintenance of a healthy resource. Greater isolation of the conservation decision from allocation pressures is the central element in successful resource maintenance programs in forestry, wildlife, water resources, and rangeland management.[3]

Market mechanisms drive allocation process. Demand commonly exceeds supply of highly valued food fish and shellfish. But while market mechanisms adjust the short-term availability, they do not protect the long-term supply. Conservation assumes that stocks will have the same social value in the future as now.[4]

All human systems are integrated wholes. Each component influences and is influenced by each other component. Systems theory provides mechanisms to divide systems into their component parts, while also acknowledging their holistic nature.[5] Management science shows how to handle complex organizations in terms of their primary subsystems. The goals of each subsystem must mesh with the overall system, but subsystems are conceptualized and operated separately. Subsystem inputs and outputs tie the whole system together. The output of the conservation determination is the allowable catch. This becomes an input for the allocation process.

Distinguishing conservation as a process separate from allocation enables clear objectives for stock maintenance to be specified. Since conservation decisions have more support when there is broad participation, the aim should be to get as many participants as possible committed to a set of conservation objectives—objectives best for maintaining the productivity of the resource.

Separation of conservation and allocation issues should: (1) achieve greater resource protection by giving conservation a broader and stronger base of support; (2) reverse the incentive structure, enabling users to see the benefits of making decisions that are in the long-term interest of the resource; (3) give all interest groups greater participation in making decisions that are in the long-term interest of the resource; and (4) reduce the costs of managing stocks.

PROBLEMS WITH SEPARATION

Successful separation of conservation and allocation requires solving a number of technical problems. For example, in a mixed-stock or mixed-species fishery, should catches be limited by the ability of the weakest species or stock to support a fishery? Some species of groundfish in a mixed-species fishery may have little or no economic value and the sustainability of the target species is the major interest. Thus, low-valued species may not be managed and as a result may become overfished, but as long as the target species is maintained this is unlikely to cause concern. On the other hand, in a salmon fishery where wild runs are depressed and it is considered prudent to preserve genetic diversity, the weakest stock could limit the take from other stocks. Such questions of social priority are common to most fisheries.

Allowable catch rates for a fishery that is being rebuilt raise another problem. Rapid rebuilding may require a particular user group to significantly reduce catches and hence add an allocation element to this conservation decision. To solve this problem, it is necessary to determine how various rebuilding rates would affect the current and projected benefits from fishing the stock. The conservation determination indicates future stock size and the catch rates to achieve it. The allocation process determines the level of current sacrifices to achieve future gains.

A problem for any migratory stock is that harvesters in different geographic locations and possibly different political jurisdictions are in direct competition for a limited resource. With salmon stocks, those near the spawning grounds are at a disadvantage relative to those who catch salmon in the ocean well before they make their trip to spawn. Conservation policies do not have to specify where salmon are to be caught. The conservation determination should simply indicate the relationship between catch levels at various points in the system. For example, more small salmon may be taken in the ocean, or fewer large salmon may be taken in the rivers. The management plan for coho salmon in the Oregon Production Index area has a sliding scale for allocation over different stock levels. When stocks are abundant, higher percentages go to commercial trollers. With low stock availability, anglers get a greater proportion. In 1986, anglers and trollers negotiated an allocation modification to this agreement, providing for a full summer angling season, and for commercial trollers to obtain the uncaught portion of the recreational allocation in years of stock abundance.

Conservation decisions approach these problems in two general ways. First, all socially important fisheries have historical patterns of allocation that evolved with development of the fishery. If there is no reason to change this historical pattern, the allocation decision assumes continuance of the same catch distribution among fishery participants. The allocation plan[6] indicates the allowable catches for each user group in proportion to past catches.

The historical pattern will not, however, be satisfactory in all cases. For social, economic, or political reasons the catch distribution in a fishery might require change. Conservation plans, however, need not set limits on specific users, but can indicate overall allowable catch and the revised relationship between the catch rates of users. This second approach gives options allowing flexibility for the allocation process to change the catch distribution among users.

Either approach, reliance on historical patterns or a revised catch distribution among users, requires detailed knowledge of the fishery. Making conservation determinations requires knowledge of the fishery's demographic, economic, and social dynamics. This knowledge has to anticipate the allocation options in order to suggest catch rate options.

The primary objective of conservation is the maintenance of a viable and healthy resource. Successful conservation planning avoids the tendency to let conservation dictate allocations or restrict user options. Many use patterns will achieve conservation goals: managers need only indicate the conservation consequences of various use patterns. But fishing must not proceed until the allocation priorities are set.

Conservation works best when the allowable catch is determined before the who, when, where, and how decisions implementing allocations. Indeed, to maintain the resource, conservation decisions must precede those on allocation. Conservation determinations should be separated conceptually, structurally, temporally, and institutionally from the allocation process. When one decision encompasses the two processes, there always will be pressure to serve the interests of more users or to encourage growth in user productivity. Thus, a tragedy of the commons occurs when individual incentives to take more fish are detrimental to the resource.

IMPLEMENTATION

One way to separate conservation and allocation within the existing structure of regional fishery management councils in the U.S. is to give the Scientific and Statistical Committees (SSCs) a greater role in formulating the conservation determination.[7] With the SSC having a greater role in developing the conservation plan, the advisory committees could have a more substantive role in formulating the management plan. An alternative is to centralize the conservation determination at the federal level, with the councils continuing to function mainly in the allocation process.

New Zealand has recently implemented an approach similar to the second alternative described above.[8] The allowable catch is determined first on resource conservation criteria at the national level. Regional organizations then allocate the quantity of fish available among groups of fishers. To increase program flexibility and ability to cope with variability in stock size, New Zealand created two licenses, one continuing, the other short-term.

Separating conservation and allocation within the Fishery Conservation and Management Act (FCMA) would require four changes. First, it is necessary to modify existing standards to distinguish conservation and allocation. Second, the presently defined concept of optimum yield needs to be redefined to associate it with the process of allocation. Third, fishery management plans become allocation plans. Add to the council functions a conservation planning requirement to determine the allowable catch available for allocation. Fourth, fishery management councils must be granted the authority to develop new organizations to carry out the allocation planning that constitutes the fishery management plan.

Many different types of organizations can allocate the allowable catch. Flexibility in creating these organizations permits them to be tailored to each fishery. As a general rule, where users are very similar, participation according to voting rights is a successful method. With a heterogeneous population of users, a shareholder system is best.

Separating conservation from allocation need not be permanent. The separation is a transitional stage to reverse the self-defeating incentives inherent in current management practices. Development of effective allocating organizations allows users to see the conservation benefits of their allocation decisions.

Giving users the task of working out allocations lets fishery management councils spend less time and money on the allocation process. Once councils prepare an effective conservation plan, users can allocate the allowable catch and prepare a management plan for council approval. Councils would be required to certify that the Fishery Management Plan (FMP) met the goals of the conservation plan and that enforcement was adequate.

WHY ALLOCATION ORGANIZATIONS?

A program giving conservation greater weight in decision making does nothing to change the incentive structure. The incentives producing a tragedy of the commons are still in place. For all users to feel the impact of their own actions on the whole, they must have some stake in the management of the resource.[9] To develop incentives for resource conservation, harvesters must collectively experience feedback as to how their individual actions affect the resource. Including users in the organizations making allocations provides this feedback.

Economists discussing open-access fisheries point out that if a profit-maximizing firm owned the resource it would not, under most circumstances, overfish and would tend to maximize profits. Open-access fisheries, on the other hand, tend to approach an economic equilibrium where total revenues equal total costs. Furthermore, the resource is typically depleted with no assurance of long-term productivity of the fish stock when economic equilibrium is approached.

Where total revenues equal total costs there is no profit. The profit-maximizing firm would not operate with no profit. Profit is maximized where total revenues exceed total costs by the greatest amount. This is where the marginal cost of adding more effort (the cost of adding the next unit of effort) equals the marginal return (the amount that this new unit of effort can earn). In static equilibrium models for fisheries this point does promote the long-term viability of the fish stock.[10]

Fishing firms are independent units. They lack perspective on what is happening to the resource. Only if one firm controlled the resource would the economic model work. Government adopts this role in fishery management, but government is not run for a profit. Governments have to meet the needs of many constituencies; if one organization operated the fishery, that organization's survival would depend on resource viability.

An organization having responsibility for both formulating and implementing a council-approved FMP is desirable. This organization would bear the costs of plan development, maintenance of contact with users, and adherence to plan provisions. The council could require enforcement mechanisms using the Coast Guard as it does now, but the allocating organization should do as much policing of its members as possible.

To involve users in allocation decisions, councils need the authority to work with organizations of users established to prepare and implement management plans. The major role of the councils would then become the evaluation of management plans according to how well they accomplish the objectives of the conservation plan.

ORGANIZATIONS WITH ALLOCATION CAPABILITIES

User associations, profit-making and not-for-profit corporations, and specially created organizations are all capable of organizing users to make allocation decisions. Depending on the nature of the group of users, each organization has advantages and disadvantages. Flexibility with respect to organizational type has to be permitted. The major function of the organization is to get users together to decide how the allowable catch will be allocated. The conservation responsibility needs to be vested with some broader entity such as a regional fishery management council or at the federal level. Each fishery requiring management has a history. The management regime used has to grow out of that history. The historic pattern of allocation among groups should not change dramatically, at least initially. In currently managed fisheries, the departure point would be the existing FMP.

Combining fishery users into an organization that collectively makes allocations has the potential for reversing the incentive structure. If decisions are made outside the group of users, they are left feeling that they are subject to someone else's authority and with no incentive to deal with the collective effects of their actions. If allocation decisions are made by the group, this should create incentives for group members to make decisions that are in the long-run best interest of the fishery. Users will become responsible to each other and to the resource, thus creating a positive incentive to make decisions that will maintain a healthy and abundant stock into the future.

Do users have the background to develop a management plan to allocate the allowable catch? Guidelines for plan preparation are explicitly presented in federal guidelines and in the National Environmental Protection Act. Plans are already in effect and the overall structure of what a plan looks like is known. No one knew how to make a management plan when the FCMA passed in 1976.

Making users responsible for decisions on allocation has a number of benefits. First, the incentive structure is reversed and the group has to confront the combined effect of each individual's behavior. Second, if the organization created is successful in making allocations, incentives of the resource users will be reversed to promote conservation as well. Third, if the users come to see the benefits of conservation, conflicts may often be reduced.

Finally, when people see the benefits of their organization, they are much more willing to pay the costs of its operation.

The primary cost of requiring user organizations to make allocation decisions is a reduction in participation and the creation of an elite group of fishers. This also occurs with governmentally operated limited entry programs. Any effort control measure reduces participation. In many fisheries, the group participation requirement would reduce the number of users. Many people who fish for the independence it gives would not like being part of a group or formalized organization.

Associations

An association of users is one type of organization enabling resource users to participate in resource allocation.[11] Association decision making is by voting. This decision-making approach operates best when members are relatively homogeneous, such as a fishery with similar gears, fishing patterns, and user populations. Voting does not work, for example, in a fishery with both recreational and commercial components. The greater number of the recreational users gives them greater decision-making impact. The same imbalance would occur in a commercial fishery composed of both trap and trawl gears.

Associations of users can successfully manage the allocation of natural resources. This occurred for a brief period in the Bay of Fundy herring fishery.[12] Among Alaskan natives, the Alaskan Eskimo Whaling Commission, Hooper Bay Waterfowl Plan, Eskimo Walrus Commission, and International Porcupine Caribou Commission, to varying degrees, allocate resources, supplement data collection, and police tribal members.[13] In addition, many of these organizations participate effectively in furthering conservation goals.

Associations of users have made compromises among interest groups and have contributed to the management process on specific issues. Here are some specific examples from Pacific Northwest fisheries: Trawl fishers recently recommended timing of catch quotas for bottom fish to maintain their market presence. A group of fixed-gear sablefish fishers voluntarily proposed to release small sablefish which could be caught legally under existing rules. Salmon anglers and trollers negotiated revised in season allocations in 1985 that were adopted by the Pacific Fishery Management Council.

Alaskan regional aquacultural associations integrate commercial, recreational, and subsistence fishing interests with those of processors and representatives of local communities. The method used for integrating these diverse interests is a landing tax to fund activities. Some of these associations have successfully gathered data, allocated added salmon produced, and policed members.[14]

In British Columbia fishers and managers developed a management system based on a set of conservation rules that determine levels of allowable catch. The catch depends on certain observations about the supply of salmon. The system of rules is fully documented, as are the management procedures. Hilborn and Luedke summarize the program, "To our knowledge it is the first opportunity commercial fishermen in British Columbia have had to participate in the setting of fishery goals and determining the rationale of the in-season management decision-making process."[15]

The association form of organization, to be successful, should encompass all fishing interests. In eastern Canada, the Bay of Fundy herring cooperative management system broke down because a competing group of fishers formed whose members were outside their effort control agreements.

Corporate organizations

Where user interests are diverse, the criticism is made that resource users cannot also be resource managers. Users do not have the time; they will make decisions in their own self-interest; and there will be too much conflict. Separating conservation from allocation offers a check against these criticisms. No matter how successful or unsuccessful the allocation process may be, conservation safeguards will be in place. The observation that resource users cannot manage their own affairs is based on the current system of fishery management which promotes rather than reduces conflict. Those managed feel alienated from the system because rules are imposed on them by an outside authority. Further, the management system promotes controversy by allowing decisions to be reversed by special interest political power. The attitude that people cannot manage their own affairs contributes to management conflicts and to

excessive costs of management. People will act cooperatively in their own self-interest when the advantages are greater than those of circumventing the process.

In profit-making corporations, ownership of shares in the organization determines participation. Shares could be in the form of quotas for catching a certain amount of fish. For commercial fisheries, the quotas should be large; if the fishery has a recreational component, the quotas should be reduced to the level of one or a few fish. A quota represents an opportunity to catch a particular quantity of fish or might be an actual catch. Participation in the corporation, then, is according to the number of shares owned.

Use of a corporate organization for allocation raises three issues. First, profit-maximizing decisions discriminate against user interests that are difficult to value monetarily. A second problem is, if the corporation profits from catching the resource, situations may develop where short-term inflation of value of the resource could stimulate overfishing. Finally, concentration of economic power is a problem, though it can be resolved if market substitutions for the fish product exist.

Not-for-profit corporate organizations can serve a diversity of user interests and create incentives to conserve the resource. A difficulty with not-for-profit corporations, however, is that they may not respond rapidly to changing market conditions if the user mix is complex. A not-for-profit corporate structure is preferred for making allocation decisions because its primary goal is organizational survival. If the not-for-profit corporation does not satisfy the needs of its members, it will cease to exist. The not-for-profit corporation has to have membership incentives that protect the resource as well as the continuance of the organization. For example, if the corporation gains strength by increasing the number of members, this will probably be detrimental to the resource. The level of participation in the corporation should come from the number of fish members catch, rather than from the number of members.

A flexible approach to creating new fishery management organizations will promote innovation in designing new organizations to conserve and allocate fishery resources. Separating the conservation and allocation responsibilities provides safeguards to protect the resource and an opportunity to experiment with new organizational forms. For example, state fishery agencies might contract for the allocation role on behalf of licensed recreational and/or commercial fishers of the state. The states of Washington, Oregon, Idaho, and California could each develop a salmon allocation plan to divide the allowable catch in their area among salmon anglers and commercial fishers. However, responsibility for conservation could remain with the regional fishery management council or transfer to a central federal authority.

SUMMARY

Fishery management encompasses the dual processes of conservation and allocation.

Conservation is a *how much* question which must be answered before making allocations to users. Viewing the resource system as a whole best achieves conservation objectives. This requires an organization with broad authority over all aspects of a resource.

Allocation answers the question of *who* uses the resource. As the diversity of users increases, so does the complexity of interests. With many different users, diverse and sometimes conflicting goals have to be accommodated. Allocation requires making judgments among many objectives, while conservation objectives are more universally accepted.

A number of organizational arrangements can accomplish the separation between conservation and allocation. These include relatively minor modifications to the FCMA, use of the existing council structure in new ways, or developing new forms of organization.

The following principles assist in achieving the conservation and allocation separation.

— Make conservation decisions giving the resource first priority. Make allocation decisions with the needs of users being primary.

— Make the conservation decision of how much before the allocation decision of who will make the catch.

— Structurally separate the conservation decision from decisions to allocate.

— All human systems are holistic and each action is related to every other action. Avoid, however, the pressure to include user-oriented allocation arguments in conservation considerations.

— Conservation requires a broad view to encompass the full range of the resource and include as many factors affecting habitat as possible. Allocation is more specific and relates to the needs of groups of users.

— Allocation seeks an optimal resolution between a number of objectives. Conservation seeks to maintain a healthy and viable resource for future generations.

ACKNOWLEDGMENTS

In preparing this manuscript I benefited from the comments and suggestions of members of the NOAA Fishery Management Study—William J. Hargis, Jr. (Chair), Richard J. Baker, FitzGerald Bemiss, James C. Cato, John P. Harville, Allen Haynie, Henry Lyman, John A. Mehos, John G. Peterson, and William E. Towell. Comments made in response to the NOAA Fishery Management Study helped in preparing this manuscript. Help of the NOAA study staff—Bruce Norman, Daphne White, Robin Tuttle, and Ann Smith—was particularly valuable.

I appreciate materials supplied by Peter Fricke, Evelyn Pinkerton, Steve Langdon, and M. Estellie Smith used in preparing this manuscript, and editorial help from Linda Varsell Smith. I am solely responsible for these ideas. This presentation does not represent an endorsement by the fishery study participants, the National Oceanic and Atmospheric Administration, or others acknowledged.

NOTES

1. This usage of "conservation" is narrower than the usual connotation of wise use of resources. The usage parallels more closely actual practice in fishery management.

Conservation was differentiated from preservation in the 19th century debates between Gifford Pinchot and John Muir. The Pinchot adherents advocated the wise use of forest lands. Muir's associates sought to prevent the commercial use of unique and pristine forest habitats.

For most of the history of fishery management biologists, who dominate management organizations, denied that they could or did allocate resources among users. The prevailing philosophy was that if good care was taken of the resource, the wise use question would solve itself. The Fishery Conservation and Management Act explicitly addresses this view. The FCMA clearly identifies that allocation occurs by both action and inaction.

As recently as 1980, the Washington Department of Fisheries argued that the state constitution prevented it from making allocations of salmon among users. This led to the federal court taking over allocations between Indian and non-Indian fishing interests in the Boldt decision.

2. Garrett Hardin popularized this phrase in an article, "The Tragedy of the Commons," *Science* 162 (December 13, 1968): 1243-1248. Economists refer to the same problem as "open-access," and H. Scott Gordon's article "The Economic Theory of a Common Property Resource: The Fishery," *Journal of Political Economy* 62(1954), pp. 124-142, is a classic statement of the open-access problem.

3. Office of Policy and Planning, National Oceanic & Atmospheric Administration. *Fishery Management— Lessons from Other Resource Management Areas* (Washington, D.C.: U.S. Department of Commerce, NOAA, Office of the Administrator, 1985); Ross W. Gorte, Eugene H. Buck, David M. Sale, and Adrienne C. Grenfell, *Limiting Access for Commercial Fish Harvesting* (Washington, D.C.: Library of Congress, Congressional Research Service, 1985), pp. 54-60.

4. Colin Clark, *Mathematical Bioeconomics* (New York: John Wiley and Sons, 1976), p. 43, discusses this point, "For this objective is based on the supposition that society is, or should be, willing to make arbitrary current sacrifices to benefit future generations, the only provision being that total long-term economic benefits must be increased thereby." See also pp. 24-64.

5. See the processes of synthesis (constructing supersystems) and projection (reducing a system into subsystems) in David O. Ellis and Fred J. Ludwig, *Systems Philosophy* (Englewood Cliffs: Prentice-Hall, Inc., 1962), pp. 11-12.

6. The term "conservation plan" means a separate process of goal setting, fishery knowledge, and allowable catch determination. The allowable catches become the basis from which the management plan, the one that makes allocations, is prepared.

7. For current council structure, see Pacific Fisheries Management Council, *Pacific Fishery Management Council function and form* (Sept. 1985), 20 p.

8. John Annala, "The Introduction of Limited Entry: The New Zealand Rock Lobster Fishery," *Marine Policy* (1984):101-108.

9. A discussion of this point is in E.A. Keen, "Common Property in Fisheries: Is Sole Ownership an Option," *Marine Policy* 7 (1983), pp. 197-212.

10. Fishery economics texts discuss this point. See Clark cited above (see note 4); James Crutchfield and Giulio Pontecorvo, *The Pacific Salmon Fisheries: A Study of Irrational Conservation* (Washington, D.C.: Resources for the Future, 1969), pp. 11-36; and Lee G. Anderson, *The Economics of Fisheries Management* (Baltimore: The Johns Hopkins University Press, 1977), pp. 22-55.

11. This is sometimes referred to as cooperative management and is used in a number of Canadian fishery management situations.

12. John F. Kearney, "The Transformation of the Bay of Fundy Herring Fisheries 1976-1978: An Experiment in Fishermen-Government Co-Management," in *Atlantic Fisheries and Coastal Communities: Fisheries Decision-Making Case Studies*, Cynthia Lamson and Arthur J. Hanson, eds. (Halifax: Dalhousie Ocean Studies Programme, 1984).

13. Evelyn W. Pinkerton, "Co-Cooperative Management of Local Fisheries: A Route to Development," Paper presented at the Fisheries Co-Management Conference, Vancouver, British Columbia, May, 1986; and Steve J. Langdon, "Alaskan Native Subsistence: Current Regulatory Regimes and Issues," Paper for Roundtable Discussions of Subsistence, Anchorage, Alaska, October 10-13, 1984.

14. Pinkerton (see n. 13 above).

15. Ray Hilborn and Wilf Luedke. "Rationalizing the Irrational: A Case Study in User Group Participation in Pacific Salmon Management," *Canadian Journal of Fish and Aquatic Sciences.*

Coastal Community Impacts of the Recreational/Commercial Allocation of Salmon in the Ocean Fisheries

Christopher N. Carter

Economist
Oregon Department of Fish and Wildlife

Hans D. Radtke

Economist
Yachats, Oregon

INTRODUCTION

On 31 May 1985, the Oregon Fish and Wildlife Commission directed its staff to compile economic and other information that would be useful in developing an explicit policy on the sport/commercial allocation of coho salmon in the ocean fisheries. This paper summarizes the progress made in developing an economic analysis of the allocation issue.

The paper contains several sections: (1) a review of recent historical data on the commercial and recreational salmon fisheries; (2) a brief explanation of value concepts; (3) an explanation of the methodology used for assessing community economic impacts; (4) an explanation of a fisheries economic assessment model which can be used to develop information on the income impacts of salmon allocation; and (5) some suggestions on the potential application of the model to allocation policy analysis.

THE COMMERCIAL AND RECREATIONAL SALMON FISHERIES

It is essential to have a general picture of the nature and size of Oregon's commercial and recreational salmon fisheries. To provide this background, we summarize historical data on the commercial and recreational salmon fisheries in this section. The primary focus of this paper (and see also Oregon Department of Fish and Wildlife 1985b) is on coho salmon; however, background information on both coho and chinook is needed to characterize the allocation problem.

For coho salmon management, the production area of direct interest to the Oregon Department of Fish and Wildlife (ODFW) and to most Oregon fishers is called the Oregon Production Index (OPI) area. This area encompasses the Columbia River and tributaries, Oregon and California coastal streams, and ocean waters south of Leadbetter Point, located near Willapa Bay, Washington. It is beyond the scope of this report to provide a detailed discussion of the historical status of coho stocks; however, Table 1 shows a summary of coho returns for 1976-84.

At the time this study was begun, the commercial/sport allocation was governed by provisions contained in the "salmon framework plan" (Pacific Fishery Management Council 1984). Separate coho allocation schemes were established for two areas—the areas north and south of Cape Falcon, Oregon. The schedule which prevailed through 1986 for the area south of Cape Falcon is the primary subject of this paper and is shown in Table 2. (Additional details are discussed in Pacific Fishery Management Council 1984.)

In this specific context, the key policy question is whether or not the allocation schedule shown in Table 2 is the "best" way to guide society's distribution of the ocean coho catch. If it is not, then what schedule and method can be produced using regulations to approach a better arrangement? In this paper we concentrate on one economic criterion for analyzing the allocation policy question. Clearly, other criteria for judging alternatives may also be extremely important.

The commercial fishery

The ocean troll salmon fishery has been one of Oregon's most important commercial fisheries for many years. More licensed vessels and fishers fish in the troll salmon fishery than in any other. In many years the harvest level revenues from this fishery exceeded revenues from any other single fishery, but there has been considerable fluctuation in both landings and revenues, and in recent years the fishery has been depressed in both the physical and economic senses. Table 3 shows the catch and harvest level revenue (ex-vessel value) trends in the ocean troll salmon fishery from 1976 through 1984.

The size of the troll salmon fleet has increased and then decreased over the course of the last 10 years in response to several factors, including variations in catch and revenues, establishment of a restricted vessel permit ("moratorium") system in 1980, and increasingly stringent regulations imposed by

Table 1. *Oregon Production Index of adult coho returns, 1976-84.*
(In 1,000s of fish)

Year	Ocean Fisheries		Inland Escapement		Total OPI	Private Hatchery Fish in Ocean Catches[a]	OPI Excluding Private Hatchery Fish[b]
	Troll	Sport	Coastal	Columbia River			
1976	2,793.5	931.1	59.8	326.3	4,110.7		
1977	632.8	392.5	10.3	87.0	1,122.6		
1978	1,051.6	503.0	10.2	297.3	1,882.1		
1979	1,005.8	319.6	41,7	264.3	1,631.4	63.0	1,568.4
1980	482.8	502.4	39.4	284.8	1,309.4	53.6	1,255.8
1981[c]	783.3	328.0	33.6	162.4	1,307.3	142.0	1,165.3
1982[c]	689.3	270.6	35.6	435.9	1,431.4	122.1	1,309.3
1983[c]	398.1	259.9	18.2	97.1	773.3	110.3	663.0
1984[c]	88.1	174.9	37.9	392.8	693.7	35.0	658.7

[a] Estimates.

[b] Adjusted OPI excludes catch of private hatchery fish and is identical to total OPI for 1970-78.

[c] Data are preliminary.

Table 2. *Schedule through the 1986 season for allocation of coho salmon south of Cape Falcon, Oregon.*

Allowable ocean harvest (In 1,000s of fish)	Commercial		Recreational	
	Number (In 1,000s)	Percentage	Number (In 1,000s)	Percentage[a]
≥ 2500	2150.0	86.0	350.0	14.0
2400	2056.8	85.7	343.2	14.3
2300	1964.2	85.4	335.8	14.6
2200	1874.4	85.2	325.6	14.8
2100	1780.8	84.8	319.2	15.2
2000	1690.0	84.5	310.0	15.5
1900	1597.9	84.1	302.1	15.9
1800	1506.6	83.7	293.4	16.3
1700	1416.1	83.3	283.9	16.7
1600	1324.8	82.8	275.2	17.2
1500	1234.5	82.3	265.5	17.7
1400	1145.2	81.8	254.8	18.2
1300	1056.0	81.2	244.0	18.8
1200	966.0	80.5	234.0	19.5
1100	876.7	79.7	223.3	20.3
1000	788.0	78.8	212.0	21.2
900	699.3	77.7	200.7	22.3
800	612.0	76.5	188.0	23.5
700	525.7	75.1	174.3	24.9
600	430.0	71.7	170.0	28.3
500	330.0	66.0	170.0	34.0
400	230.0	57.5	170.0	42.5
300	130.0	43.3	170.0	85.0
200	30.0	15.0	170.0	85.0
≤ 100	[b]	[b]	100.0	100.0

[a] For allowable coho harvests of 700,000 and above, the allocation shall be interpolated linearly between the numbers shown.

[b] Incidental coho allowance associated with directed chinook fishery would be deducted from recreational catch. Incidental allowance could be in the form of estimated hooking mortality or actual landing allowance.

government. Table 4 indicates the overall trend in numbers of vessels active in the troll salmon fishery from 1976 through 1984.

The Oregon troll salmon fleet is not composed of a number of average, homogeneous firms. There is great diversity in every descriptive characteristic of the fleet, including vessel size, owner motivation, activities in other fisheries or states, and catch distribution by vessel. Table 5 provides summary information that indicates that roughly 12 percent

Table 3. *Ocean troll salmon landings in Oregon with ex-vessel values, 1976-84.*

Year	Number of Deliveries	Coho				Chinook				Total Value $
		Fish	Pounds	Price $	Value $	Fish	Pounds[a]	Price[b] $	Value $	
1976	75,800	1,827,000	9,061,200	1.26	11,458,000	184,300	1,921,600	1.77	3,410,000	14,868,000
1977	85,100	446,100	2,640,800	1.34	3,546,000	340,000	3,464,900	2.17	7,938,000	11,484,000
1978	45,700	611,600	2,779,000	1.35	3,756,000	191,500	1,893,600	1.89	3,584,000	7,340,000
1979	43,600	714,600	4,586,300	2.26	10,350,000	245,500	2,580,000	2.57	6,639,000	16,988,000
1980	29,900	383,300	2,190,200	1.34	2,926,000	209,400	2,171,500	2.42	5,259,000	8,185,000
1981	35,100	620,300	3,324,300	1.66	5,534,000	160,400	1,573,400	2.57	4,039,000	9,570,000
1982	26,500	521,900	2,708,400	1.40	3,801,000	232,800	2,351,300	2.59	6,094,000	9,895,000
1983	17,400	319,800	1,098,000	0.96	1,052,000	79,600	654,800	1.90	1,244,000	2,296,000
1984	5,400	14,000	71,000	1.66	118,000	63,600	539,300	2.74	1,477,000	1,595,000

[a] Dressed weight

[b] Average nominal price per pound dressed weight

Table 4. *Number of vessels landing troll-caught salmon in Oregon, 1976-84.*

Year	Number of Vessels[a]	
1976	2,770	
1977	3,108	
1978	3,158	
1979	3,114	
1980[b]	3,875	(4,314)
1981	3,615	(3,926)
1982	3,269	(3,646)
1983	2,948	(3,437)
1984	771[c]	(3,197)

[a] Number of vessels holding permits for troll salmon fishing under the state license moratorium, but not necessarily landing salmon in Oregon are noted in parentheses.

[b] The establishment of a restricted vessel permit system (with liberal initial qualification criteria) drew a number of previously active vessels back into the fishery in 1980.

[c] Vessels were not required to land one salmon in 1984 to be eligible to renew permits in 1985. The Oregon Commercial Fishing Vessel Permit Board waived this requirement because of closure of the coho fishery south of Cape Falcon.

Table 5. *Number of vessels landing 50 percent and 90 percent by weight of total Oregon salmon troll catch, 1976-84.*

Year	Total vessels	Landing 50%		Landing 90%	
		Number	Percentage	Number	Percentage
1976	2,770	453	16.4	1,460	52.7
1977	3,108	473	15.2	1,597	51.4
1978	3,157	446	14.1	1,576	49.9
1979	3,114	423	13.6	1,449	46.5
1980	3,875	372	9.6	1,375	35.5
1981	3,615	420	11.6	1,391	38.5
1982	3,269	359	11.0	1,249	38.2
1983	2,951	294	10.0	1,082	36.7
1984	771	88	11.4	333	43.2

Note: Data include licensed (permitted for 1980-84) and properly identified vessels only.

of the vessels land 50 percent of the total poundage of salmon, while 40 percent of the vessels land 90 percent. Thus, the majority of salmon are caught by a minority of vessels, perhaps no more than one-third of the entire fleet.

Regulations governing Oregon troll salmon fishing seasons have become generally more restrictive over the years. Prior to 1948 there were no seasonal restrictions at all. From 1949 through 1975 the length of the coho fishing season was 139 days; the chinook season lasted roughly two months longer. Table 6 shows season lengths for the area south of Cape Falcon for 1976-84.

Table 6. *Oregon troll salmon fishing seasons, south of Cape Falcon, 1976-84.*

Year	Length of Fishing Season (Days)	
	Coho	Chinook[a]
1976	139	184
1977	139	184
1978	139	184
1979	65	154
1980	56	155[b]
1981	55	154
1982	12[b]	169[b]
1983	25-60[c]	103-144[d]
1984	0	129[b]

Note: Strictly speaking, the area dividing line has been Cape Falcon only since 1978. For 1976 and 1977 data are for the area south of Tillamook Head.

[a] Excludes late special south coast troll seasons off the Chetco and Elk rivers.

[b] Cape Falcon to Cape Blanco.

[c] Cape Falcon to Cape Kiwanda — 35 days; Cape Kiwanda to Heceta Head — 60 days; Heceta Head to Oregon border — 25 days

[d] Cape Falcon to Heceta Head — 103 days; Heceta Head to Cape Blanco — 144 days; Cape Blanco to Oregon border — 114 days

The recreational fishery

The ocean recreational salmon fishery is among Oregon's most important recreational fisheries. According to Oregon's last survey[3] of licensed anglers (Survey Research Center 1978), the number of salmon fishing days (21.6 percent of total effort) exceeded use levels in any other sport fishery except for the inland trout fishery. The estimated effort and catch in the Oregon ocean salmon recreational fishery from 1976-84 are shown in Table 7. Coho catch data south of Cape Falcon for that period is shown in Table 8.

The magnitude and species composition of the recreational catch varies from area to area (Table 9). Although the coho catch exceeds the chinook catch in most Oregon areas, the chinook catch in the Brookings area has been roughly as great as the coho catch since 1981. The California ocean recreational salmon catch is generally dominated by catches of chinook.

The time distribution of catch and effort varies from year to year and area to area depending on stock abundance and distribution and regulations. Table 10 shows the distribution of effort (angler days) for the aggregate of Oregon areas south of Cape Falcon for the years 1979-84. The effects of

Table 7. *Ocean salmon recreational effort and catch off Oregon, 1976-84.*

Year	Salmon Trips	Total Trips	Catch (No. of Fish)			Salmon Per Angler Trip[b]
			Coho	Chinook	Total[a]	
1976	NA	538,400	501,300	79,300	580,600	1.08
1977	NA	404,500	195,300	61,400	260,700	0.64
1978	NA	403,700	259,800	22,800	282,600	0.70
1979	301,300	341,800	180,800	20,900	202,300	0.67
1980	331,400	362,000	325,800	19,000	344,900	1.04
1981	311,000	346,700	199,800	29,200	230,600	0.74
1982	226,000	249,700	175,100	38,700	213,800	0.95
1983	226,000	261,500	146,900	24,700	171,700	0.76
1984	153,100	203,600	123,300	17,000	140,300	0.92

[a] Includes the catch of a few pink salmon in odd years.

[b] Fish/angler computed based on total angler trips prior to 1979.

an extremely early season closure in 1982 and a restrictive season in 1984 are apparent from the data. Ocean recreational coho catch distribution by month is displayed in Table 11, and presents a similar picture.

Seasons and bag limits in the ocean recreational salmon fishery have changed considerably over the years. An abbreviated summary of key information on sport seasons for coho during the period 1976-84 is shown in Table 12. Prior to 1976 there was no limitation on the length of the sport season.

The distribution of angler effort and salmon catch by mode (type of boat) has varied over the years (Figure 1). The majority of trips and catch have been taken in private (pleasure) boats. The percentage of trips taken in charter boats has declined especially over the last 4 years. The numbers of trips and catch across species and modes have generally declined of late as indicated by Table 13. Oregon began issuing licenses specifically for

Table 8. *Recreational catch of coho salmon south of Cape Falcon, Oregon, 1976-84.*

Year	Oregon Catch South of Cape Falcon[a]	California Catch	Total
1976	384,700	57,900	442,600
1977	139,900	14,200	154,100
1978	199,800	44,400	244,200
1979	142,900	16,500	159,400
1980	270,200	22,100	292,300
1981	145,600	9,700	155,300
1982	139,600	24,600	164,200
1983	109,600	26,900	136,500
1984	112,500	18,400	130,900

[a] Calculated as the sum of catches in all Oregon areas except the Columbia River Area.

charter boats in 1980 (Table 14). The trend in license sales to Oregon residents is downward, probably reflecting depressed fishery conditions. Licenses issued to Washington residents are primarily for vessels operating out of southwest Washington ports which may find advantage in fishing in Oregon waters at certain times.

ECONOMIC VALUE CONCEPTS AND SALMON ALLOCATION
Meaning of value

More often than not when we are asked about the value of salmon there is a presumption that an all-encompassing measure of value can be developed and even applied across the board on a per fish basis. The usual question that people want answered is, "What is the value of (a) salmon to Oregon's economy?" Unfortunately, there is no correct way to provide a single, simplistic answer to this question. A detailed discussion of value is beyond the scope of this report; however, we here try to summarize some of the important aspects of salmon valuation. The reader is referred to Rettig (1984) for a more complete discussion.

In the analysis of policy alternatives we are interested in the value of products and activities which would increase or decline as a result of our choice among the alternatives. It may not be particularly helpful to try to value the salmon themselves since that value varies over time, from area to area, and by use.

Value estimates are most helpful to decision makers when (1) the specific measure of value estimated is clearly defined and understood; (2) the value estimates for all alternatives are expressed in

Table 9A. Ocean recreational coho catch in Oregon areas south of Cape Falcon, 1976-84.

Year	Tillamook Area[a]	Newport Area[b]	Coos Bay Area[c]	Brookings Area[d]	Sum Over Areas
1976	50,000	118,100	164,800	51,800	384,700
1977	15,600	34,000	76,300	14,000	139,900
1978	8,500	61,400	82,800	47,100	199,800
1979	9,400	36,300	79,000	18,200	142,900
1980	28,900	72,400	135,900	33,000	270,200
1981	17,800	61,900	57,600	8,300	145,600
1982	23,100	44,000	55,600	16,900	139,600
1983	8,800	21,800	62,700	16,300	109,600
1984	20,300	41,200	39,400	11,500	112,400

Table 9B. Ocean recreational chinook catch in Oregon areas south of Cape Falcon, 1976-84.

Year	Tillamook Area[a]	Newport Area[b]	Coos Bay Area[c]	Brookings Area[d]	Sum Over Areas
1976	2,300	4,600	14,600	13,200	34,700
1977	1,500	2,600	22,700	11,800	38,600
1978	800	2,100	4,800	7,300	15,000
1979	1,000	1,400	4,500	6,400	13,300
1980	1,600	1,800	5,300	4,800	13,500
1981	1,900	2,400	4,500	8,900	17,700
1982	1,400	3,500	10,100	15,500	30,500
1983	700	1,600	6,600	12,400	21,300
1984	1,100	2,000	4,900	9,000	17,000

[a] Includes Garibaldi and Pacific City.
[b] Includes Depoe Bay and Newport.
[c] Includes Florence, Winchester Bay and Coos Bay.
[d] Includes Gold Beach and Brookings.

comparable units; and *(3)* potential gains are expressed in terms comparable to potential losses. Furthermore, it is not the total value of affected activities which is of interest, but the change in value resulting from a choice among policy alternatives. The most important questions in valuing alternatives are: Who is involved? What are the objectives? What activities are being changed? What is the causal chain from action chosen to objectives affected? What scales are to be used to measure benefits (gains) and costs (losses)? How can value information be estimated? What assessment is warranted from the information collected? Thus, "valuation is a process which must be conducted anew for each problem" (Rettig 1984).

Valuation and salmon allocation

There are no measures of value per commercially caught salmon or sport caught salmon that imply that allocation decisions should be biased in one particular direction under all circumstances. It is common to cite alleged measures of the gross values of salmon as justification for a particular allocation. Thus, for the commercial fishery, value is often associated with total receipts at one or more market levels. For the recreational fishery, value is often equated with anglers' expenditures (costs) on recreational fishing activities.

Although there are several types of values that may be used in an analysis of alternatives, two types of economic value measures are commonly applied to changes in activity levels associated with allocation alternatives. The first measure, which is appropriate for cost-benefit analysis, is *net economic value*. Roughly speaking, this is the difference between the gross value of an economic activity and the costs (properly defined and measured) of carrying out that activity. The second measure is the *impact on community income*. By income we mean the income people receive in the form of wages, salaries, and proprietary income (profits). It is this latter measure which is addressed and evaluated in this paper. Community impacts seem to be the values in which state policy makers and many members of the public have the greatest interest.

Table 10. Monthly distribution of recreational fishing effort for Oregon ocean areas south of Cape Falcon, 1979-84.
(In Angler Days)

Year	May	June	July	Aug.	Sept.	Oct.	Nov.	Season Total
1979	8,100	37,800	115,000	87,000	4,400	5,000	900	258,200
1980	4,600	70,600	126,300	65,000	15,300	2,900	400	285,100
1981	12,800	40,800	98,100	90,200	21,800	2,300	300	266,300
1982	700	27,200	154,800	10,200	3,000	2,600	300	198,800
1983	800	28,800	95,100	43,600	21,700	5,000	100	195,100
1984	0	0	72,900	64,700	5,300	3,100	0	146,000

Table 11. *Monthly distribution of recreational coho catch for Oregon ocean areas south of Cape Falcon, 1979-84.*
(In numbers of fish)

Year	May	June	July	Aug.	Sept.	Oct.	Nov.	Season Total
1979	5,300	29,100	72,500	25,900	100	0	0	142,900
1980	4,200	96,900	136,200	29,500	3,200	0	0	270,100
1981	7,500	18,000	52,700	60,500	6,900	0	0	145,600
1982	<50	18,800	120,800	0	0	0	0	139,600
1983	500	26,400	56,700	16,700	9,300	0	0	109,600
1984	0	0	61,000	50,600	900	0	0	112,500

ECONOMIC ANALYSIS OF COMMUNITY IMPACTS—METHODOLOGY

People interested in economic stability or economic development in coastal communities are often interested in estimating the impacts of economic changes (such as plant openings or closings, changes in available timber or fish for harvest, etc.) on employment, business activity, income, or public service demands.

Input/output models

Economic input/output (I/O) models are often used to estimate the impact of resource changes or to calculate the contributions of an industry to the local economy. The basic premise of the I/O framework is that each industry sells its output to other industries and final consumers and in turn purchases goods and services from other industries and primary factors of production. Therefore, the economic performance of each industry can be

determined by changes in both final demand and the specific interindustry relationships.

I/O models can be constructed using surveys of a regional economy. The disadvantages of the survey model approach are its complexity and high cost. Construction of a survey data I/O model involves obtaining data on the sectoral distribution of local purchases and sales to final demand of every sector of the economy, and on the imports purchased and exports sold by each sector. The amount of data needed to construct an I/O table and the associated time, cost, and technical skill requirements are enormous.

Another approach (taken in this paper) uses secondary data to construct estimates of local economic activity. The U.S. Forest Service has developed a computer program called IMPLAN (Siverts et al. 1983) which can be used to construct county or multicounty I/O models for any region in the U.S. The regional I/O models used by the Forest Service are derived from technical coefficients of a

Table 12. *Summary of ocean recreational coho seasons and regulations for Oregon waters south of Cape Falcon, 1976-84.*

Year	Opening Date	Closing Date	No. Days	Bag Limit
1976	May 1	Dec 31	246	3
1977	Apr 30	Oct 31	185	3
1978	Apr 29	Oct 31	186	3
1979	May 12	Sep 3	115	2
1980	May 10	Sep 14	128	3/2[a]
1981	May 15	Sep 20	129	2
1982	May 29	Aug 1	65	2
1983[b]	June 18	Sep 18	93	2
1984[c]	July 9	Aug 7	30	2

Note: Prior to 1978, the area dividing line was Tillamook Head.

[a] Bag limit 3 per day, reduced to 2 per day July 16.

[b] Season shown is for Cape Falcon to Cape Blanco. The season from Cape Blanco to the Oregon-California border was May 28-Sept. 8 (114 days) for coho.

[c] Season was open from Aug. 25-Sept. 3 (10 days) until federal preemption.

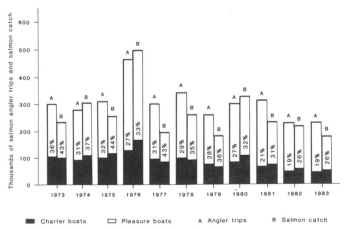

Figure 1. *Distribution of ocean salmon angler trips and catch by mode, 1973-83.*

Table 13. Oregon ocean recreational catch and effort by boat type, 1979-84.

Year	Effort (Angler Trips)		Chinook Catch (No. of fish)		Coho Catch (No. of fish)	
	Charter	Pleasure	Charter	Pleasure	Charter	Pleasure
1979	73,700	187,700	5,400	13,300	59,800	101,800
1980	79,000	218,900	5,100	11,900	98,300	207,500
1981	65,400	242,600	6,600	22,200	64,500	135,300
1982	43,300	182,700	8,200	30,600	48,500	126,700
1983	41,900	184,100	4,700	20,000	39,700	107,200
1984[a]	24,300	128,700	2,200	14,800	27,300	96,100
1979-1984 average	54,600	190,783	5,367	18,800	56,400	129,100

Note: Salmon data from surveyed ports only. For 1979-80 this includes: Astoria, Garibaldi, Depoe Bay, Newport, Winchester Bay, Coos Bay, Gold Beach, and Brookings. In 1981-84, Pacific City and Florence were also surveyed.

[a] Preliminary.

national I/O model and localized estimates of total gross outputs by sectors. The computer program (IMPLAN) adjusts the national level data to fit the economic composition and estimated trade balance of a chosen region.

Measuring the importance of local economic activity

One way of measuring the importance of a particular economic activity is to look at the amount of goods and services it sells and buys outside the local economy. A local economy has exports and imports similar to state or national exports and imports. Seafood harvested and processed in Grays Harbor or Astoria and shipped to Los Angeles is an export that benefits the local economy. The windsurfer from Seattle and the beachcomber from Boise bring money to the Hood River and Crescent City economies; these recreation activities are also exports because they bring in outside money. Exports from the local economy stimulate local economic activity.

Table 14. Number of charter boats licensed in Oregon, 1980-84.

Year	Total Number	By Oregon Residents	By Washington Residents	By Residents of Other States
1980	194	192	2	0
1981	248	213	34	1
1982	253	212	40	1
1983	255	206	47	2
1984[a]	218	185	31	2

[a] Preliminary.

However, the money brought into a local economy does not all stay in the local economy. This is particularly true for the smaller coastal economies which are far from economically self-sufficient. Many of the goods and services consumed in the local economy must be brought in from outside. They are the imports to the local economy. The money that flows out of the local economy to pay for these imports is referred to as "leakage."

Basic sectors

Since imports take money out of the economy, it is important for these smaller coastal economies to have some exporting sectors. In I/O jargon, these are called "basic sectors." The dollars brought in by basic exporting sectors stimulate a local economy by originating the multiplier process. When people talk about a change in the economic base of an area, they are referring to a change in basic business sector.

Sectors other than basic sectors generally do not generate new dollars, but rather operate on the circulation of dollars already present in the economy. Therefore, nonbasic sectors do not initiate a multiplier effect themselves, but instead contribute to the multiplier effect of basic sectors by preventing leakage. For the coastal communities, the basic sectors are generally resource-based. Examples (not necessarily in order of importance) are: basic sector— fish harvesting and processing, logging and timber processing, tourism and recreation, transfer payments; nonbasic sector— medical services, movie theaters, grocery stores, banking services. Transfer payments include social security payments, retirement payments, and nonlocal government salaries.

Calculating multipliers and coefficients

Output (sales) multipliers.[1] How is the effect of a dollar of export sales multiplied in a local economy? Suppose a county's fishing industry increases export sales by $1,000. If the economy has an *output multiplier* (derived from an I/O model) of 2.49, total business sales through the county are expected to increase by a total of $2,490 as a result of the $1,000 increase in exports and the $1,490 in local sales generated by these exports. (The 2.49 is used as an example only. The actual output multiplier may be different.)

Figure 2 demonstrates how local respending of the export payment by businesses and households creates this multiplier effect. The process begins when a dollar enters the local economy, in this case as the result of an export sale (column A). The dollar is respent by the exporting firm in order to purchase inputs (goods, services, labor, taxes, etc.) to meet the increased export demand (column B). Sixty cents of the dollar are received by local businesses and households, but $0.40 leaks out in the form of nonlocal purchases. Thus, in addition to the initial dollar, business respending has generated an additional $0.60 of business activity within the economy. Of the $0.60 that is locally received, $0.38 is respent within the county, and the rest leaks out (column C). This process continues until the amount remaining in the local economy is negligible (columns D, E, F). Thus, greater leakage at any round of respending leads to a smaller multiplier.

In order to determine the total multiplier value, the initial dollar is added to the sum of the local respending. In this example, the multiplier equals 2.38 ($1.00 + 0.60 + 0.38 + 0.20 + 0.12 + .08 + etc.). Thus, $2.38 of local business activity will be generated for $1 that enters the local economy. The same process can be used to explain a decrease in export sales.

---

[1] The multiplier presented in this example is an output (sales) multiplier, measuring the total change in local sales generated by a $1 increase in export sales. While output multipliers are useful in describing the interrelationships between business sectors, they do not adequately describe the amount of income or employment generated locally by specific business activities.

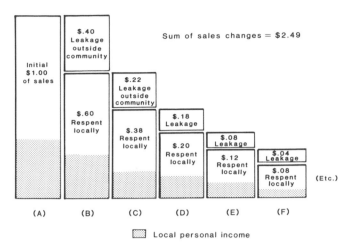

Figure 2. Output (sales) multiplier.

Calculating income coefficients. A more useful measurement of the contribution of a sector's activity is the amount of local personal income that is directly and indirectly generated from an increase in sales (Figure 3). Local personal income generated is the shaded part of the output described in Figure 2. The "local personal income coefficient" measures the income generated as a result of a change in sales. A personal income coefficient of 0.77 is used as an example in Figure 2. In the first round of export sales, $0.32 of local personal income is generated. The other $0.68 in the initial round goes to purchase supplies and services from other industries. These industries also create wages, salaries, and profits. As these sales work through the economy, a total of $0.77 of personal income is

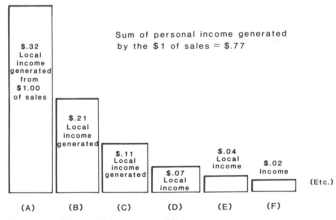

Figure 3. Personal income coefficient.

generated for every dollar of increase in sales. (Once again, this figure is an exampe only; the actual personal income coefficient may be different.) The size of the personal income coefficient is largely determined by the amount of personal income generated by the first round.

In an industry that is very labor intensive, the output (sales) multiplier may not be very large but the income coefficient may be above average. On the other hand, if the industry goes through several transactions but is not very labor intensive throughout the process, the output (sales) multipliers may be large and the income coefficient small. The output (sales) multiplier calculates how much money is "stirred up" in the economy, but it does not mean that someone in the local area is making a wage or profit from this money. The differences between output multipliers and income coefficients are often confused, leading to misuse. People, especially decision makers, need to know and understand what type of multiplier or coefficient is being used in the economic assessment of proposed changes.

The impacts estimated in this paper are effects on total personal income, the amount that is retained as household income (salaries, wages, and proprietary income). Because many jobs in the fishing industry are not full-time, an employment figure could be misleading. A full-time equivalent employment figure can be calculated by dividing the total personal income figure by a representative annual personal income average.

FISHERIES ECONOMIC ASSESSMENT MODEL FOR COMMUNITIES ON THE OREGON COAST

Input/output models have been constructed for the Oregon coastal counties with the use of the U.S. Forest Service IMPLAN model.

On the commercial side, representative budgets from the fish harvesting sector (Table 15) and the fish processing sector (Table 16), as well as a price and cost structure for processing (Table 17) are used to estimate the impacts of changes. On the recreational side, a charter operator budget (Table 18) and recreational fisher destination expenditures (Table 19) provide the basic data. The individual expenditure categories are used to estimate the total community income impacts. The commercial fisheries data were developed by Hans Radtke and

Table 15. Budget for salmon troller.

Type of Vessel: 42 foot troller
Market value $24,000[a]
1 crew and skipper[b]
Gross revenue $35,000
12% loan or investment return

	Net Cash Flow	Percentage of Total Revenues
	$	
Revenue	35,000	
Less Expenses:		
Variable Expenses:		
Repair work	1,440	0.0410
Gear replacement	2,400	0.0684
Fuel and lubricants	3,590	0.1026
Food and supplies	1,795	0.0513
Ice and bait	360	0.0103
Dues and fees	240	0.0068
Transportation	880	0.0252
Miscellaneous	880	0.2520
Crew shares (39% of gross)[b]	13,650	0.3900
Total Variable Costs	25,235	0.7208
Contribution Margin		
Fixed Expenses:		
Insurance	480	0.0137
Moorage	720	0.0205
Interest expense[a]	2,875	0.0821
Depreciation[a]		
Licenses	360	0.0103
Miscellaneous	25	0.0007
Total Fixed Expenses	4,460	0.1273
Net Return	5,305	0.1519

[a] It is assumed that the total amount of the purchase price of the boat is borrowed; if the boat owner's money is used, this is considered a return on his or her investment. Depreciation is frequently viewed as the value of principal payments; depreciation of boat and equipment is taken over a 10-year period. Market value of the boat may also include market value of boat license or fishing contract.

[b] Crew share formula and the number of crew will vary from boat to boat and from fishery to fishery; the shares vary from a percentage of gross revenues to formulas that may include deductions for food, fuel, employment tax, etc. The payment to the skipper is part of crew shares; the skipper may also be the boat owner.

William Jensen in connection with a project to develop a fisheries economic assessment model for the West Coast Fisheries Development Foundation. The budgets for recreational charter boats and recreational private boat fishers were developed from Crutchfield and Schelle (1979).

Total impacts on community income resulting from changes in final demand or output depend on the size of the direct, indirect, and induced income coefficients for the sector that is affected by the

Table 16. Representative budget for fish processor.

	Taxable Income	Net Cash Flow
Revenue	10,061,077	10,061,077
Less Expenses:		
Variable Expenses:		
Raw Product Cost[a]	6,961,057	6,961,057
Direct (Processing) Labor	1,493,677	1,493,677
General Costs and Packaging[b]	288,632.5	288,632.5
Other Variable Expenses	0	0
Bad Debt Expense	50,305.38	50,305.38
Total Variable Expenses	8,793,672	8,793,672
Contribution Margin	1,267,405	1,267,405
Fixed Expenses:		
Administrative Salaries[c]	330,000	330,000
Maintenance and Repair	75,000	75,000
Utilities	45,000	45,000
Telephone	40,000	40,000
Insurance	25,000	25,000
Taxes	25,000	25,000
Supplies	30,000	30,000
Miscellaneous	30,000	30,000
Depreciation[d]	350,000	0
Interest Expense[d]	420,000	420,000
Total Fixed Expenses	1,370,000	1,020,000
Operating Income	– 102,595	247,405

Note: Business is assumed to be mixed, large-size fish processor, with 120-200 employees and a market value of $3,500,000.

[a] Includes fish tax.

[b] Includes general costs of processing, such as equipment rentals, can costs, and chemical additives. Costs of packaging are normally borne by the buyer.

[c] Total personnel = 11.

[d] Assume 12 percent interest and 10 year depreciation—actual may be more or less.

change. To utilize the total personal income coefficients, the sector's total gross output change, adjusted for trade margins as appropriate, is multiplied by the total personal income coefficient (derived from IMPLAN). As an example, the coefficients used for the commercial salmon harvester (Clatsop County) are displayed in Table 20. An explanation of these impacts is as follows: As one additional dollar is spent on boat building and repairing in Clatsop County, $.3016 (direct income coefficient)

is retained in the local area as income (wages, salaries, and profits) in this sector. The other $.6984 is spent on local supplies and on supplies from outside the area. The amount that leaves the area is considered an import which does not generate local income. The purchases made from suppliers in the area generate an additional $.0694 of income (indirect income coefficient) in the local area. As the workers and proprietors in the boat building and repairing sector and their suppliers spend their income on general consumer items, they generate an additional $.1401 (induced income coefficient) of income in the local economy. The total local income impact is $.5111 (total income coefficient). This means that for every dollar spent on local boat building or repair, $.5111 of local income is generated.

Estimating local income impacts

The type of expense, the percent of total expenditure category, and the appropriate total income coefficient are used to estimate local income impacts. For Astoria, for example, a troll fisher spends 4.1 percent of revenues from salmon fishing on vessel and engine repair. These expenditures create $.021 of local income (.041 [expenditures] × .5111 [Astoria vessel and engine repair total income coefficient] = $.021 total local income generated from vessel and engine repair expenditures).

All of the remaining variable expenditures are estimated to create $.625 of local personal income. This $.625 (for Astoria) is added to the initial change in return to households (crew share and interest and net return) to estimate the total local income effect of each dollar of additional revenues generated in the harvesting sector of the fishing industry. Similar calculations are made to estimate the contribution to local personal income of the processing sector and for private boats and charter boats.

Tables 21 and 22 present estimates of the economic impacts on Oregon coastal communities for commercial salmon and recreational salmon fisheries. These basic estimates can be changed as different assumptions are made about the commercial or recreational fisheries (e.g., changes in prices, average weights, demographic factors, or behavioral assumptions). The estimates presented in Table 21 assume 1985 prices and average weights.

Table 17. *Assumed price and cost structure for west coast salmon.*

Species and fishery	Landed price of raw product $	Yield of processed product %	Raw product cost $	Processing labor cost $	Other processing costs $	Bad debt expense $	Variable cost of processed product $	Sales price of processed product $	Contributional margin of processed product $
Coho (troll)	1.51	97.5	1.54872	.15	.02	.0118	1.730518	2.36	.6294821
Chinook (troll)	2.48	97.5	2.54359	.15	.02	.0178	2.731390	3.56	.8286103
Pink (troll)	.65	97.5	.666667	.15	.02	.00625	.8429167	1.25	.4070833
Coho (gillnet)	.83	80	1.0375	.25	.02	.01075	1.31825	2.15	.83175
Chinook (gillnet)	1.04	80	1.3	.25	.02	.0145	1.5845	2.9	1.3155
Tule Chinook (gillnet)	.31	80	.3875	.25	.02	.0045	.662	.9	.238
Spring Chinook (gillnet)	3.00	80	3.75	.25	.02	.02375	4.04375	4.75	.70625
Pink (gillnet)	.45	80	.5625	.25	.02	.005	.8375	.1	.1625
Sockeye (gillnet)	1.14	80	1.425	.25	.02	.0115	1.7065	2.3	.5935

APPLICATION OF THE FISHERIES ECONOMIC ASSESSMENT MODEL TO COHO ALLOCATION POLICY

In this section we discuss the potential application of information developed in the preceding pages to the design and evaluation of coho allocation policy. The purpose is not to develop and assess alternatives, but to suggest key issues that should be considered in that process.

Probably the first issue that needs to be considered is the appropriate goals and objectives of allocation policy. Some examples of objectives are: (1) improving coastal community income and employ-

ment levels; (2) maximizing the net economic value of the salmon fisheries; and (3) maintaining existing lifestyles and the character of coastal communities. The appropriate model or gauge for determining success depends upon the objective or weighted combination of objectives chosen. The I/O model approach for assessing the effects of alternative policies or coastal community income will be most helpful when the first objective is important.

Relationship between recreational fish allocation and use

One of the most important things to understand concerns comparison of the estimates presented in Table 21 for commercial fishing to those in Table 22 for recreational fishing. The commercial estimates are for income impacts *per fish.* The recreational estimates are for income impacts *per recreational fishing day.*

Historical data (Table 7) suggest that each angler-day of recreational fishing produces, on the average, roughly one fish. It is tempting to conclude that each fish caught by a recreational fisher produces a community income impact of at least $36. Further, it is tempting to compare the $36 figure to the $16-18 income impact figures for a commercially harvested and processed coho. It would appear that an unambiguous economic case for reallocation from commercial to sport fishing has been made. This would be an incorrect inference.

Table 18. *Charter boat operator expenses.*

Category	Percent of Total	$/Day
Crew wages	5.0	1.56
Imputed skipper salary	31.3	9.78
Fuel	9.4	2.94
Moorage	1.4	0.44
Maintenance & repair	7.3	2.28
Insurance	4.1	1.28
Booking commission & fees	10.9	3.41
Other	1.7	0.53
Taxes, fees, license, etc.	7.5	2.35
Residual Gross (Profit or interest payment)	21.4	6.69
Total	100.0	31.26

Source: Based on Crutchfield and Schelle (1979).

Note: Data are adjusted to 1985 dollars, using the GNP price deflation.

Suppose a large number of coho is reallocated from the commercial to the recreational fishery and at the same time the daily bag limit is increased to six fish. As a result of the increased bag limit, suppose the average catch per day increases to three coho but no more money has entered the local economy. In effect, the income impact per average coho caught by a recreational fisher is reduced to $36/day ÷ 3 fish/day = $12 per sport caught fish. This $12 impact is not greater than the $16-18 per commercial coho impact and hence the reallocation scheme does not clearly produce improved coastal community income.

This hypothetical example has some implications for the magnitude of reallocation and the structure of recreational regulations which may accompany it. Simply put, the community income impacts of reallocation depend on the effect on angler effort and tourism induced as a result of the reallocation. It is the additional effort and resulting expenditures in the coastal communities which can produce positive and significant impacts. Several policy variables can influence the results:

— Total number of fish allocated to the recreational fishery. (Does the number exceed the amount that can be utilized under reasonable bag limits and season lengths?)

— Daily and weekly bag limits. (Will a daily bag limit of one fish stimulate effort? Will a large bag limit use the total sport allocation without a proportional increase in angler use and tourism?)

— Timing of seasons. (When are the fish available to various geographical areas? Will the capacity of tourist facilities and publicly provided services—the local infrastructure—at a particular port be sufficient to support additional tourist fishers? In July? In May or September?)

Good specific choices among alternatives for each policy variable could lead to longer recreational seasons, and give both the potential recreational "customer" and the supporting industry stability and the ability to plan ahead. These choices should also be made with some understanding of the likely reductions in income from commercial fishing. In this way a balanced set of regulations might increase overall community income and, at the same time, not impose undue hardships (losses

Table 19. Destination expenditures of ocean recreational salmon fisher.

	Expenditures Per Angler Day	
	Charter boat Angler $	Private boat Angler $
Restaurants	10.83	10.83
Groceries	5.26	5.26
Camping, etc.	3.02	3.02
Lodging	5.94	5.94
Boat/motor rental fees	NA	0.22
Boat landing fees	NA	1.87
Gas for boat	NA	14.48
Charter boat fees	31.26	NA
Miscellaneous	4.30	4.30
Total	60.61	45.92

Source: Based on Crutchfield and Schelle (1979).
Note: Data are adjusted to 1985 dollars using the GNP price deflator.

of income) on the commercial sectors and their dependent community. While any change in allocation can have positive or negative impacts on specific businesses, it is important that the general community economies be improved by these changes.

Winners and losers

Any major reallocation will produce "winners" and "losers." A reallocation of coho from commercial to sport fishing would obviously result in reduced incomes to sectors related to commercial fishing. Those commercial fishers who concentrate

Table 20. Astoria area (Clatsop County) income coefficients for the commercial fisheries economic assessment model.

	Personal	IMPLAN Income Coefficients		
	Direct	Indirect	Induced	Total
Variable Expenditures: *Salmon Harvester*				
Personal expenditures (wages, salaries, profit)	.3262	.0454	.2512	.6229
Repair work	.3016	.0694	.1401	.5111
Gear replacement	.2934	.0414	.2885	.6233
Fuel & oil	.3601	.0524	.3035	.7160
Ice & bait	.5015	.0877	.3173	.9066
Food & supplies	.2842	.0627	.2603	.6071
Transportation	.4198	.0512	.1909	.6619
Dues & fees	.3072	.0483	.7700	.5932
Miscellaneous	.6560	.0852	.5044	1.2456

Table 21. Total local economic (income) impacts of ocean salmon commercial fishing.
(In $ per fish)

	Astoria	Newport	Coos Bay
Chinook	$45.21	$41.08	$42.68
Coho	18.62	16.48	16.30

Notes: Chinook are estimated to average 9.4 pounds landed weight at $2.48 landed price per pound. Coho are estimated to average 5.8 pounds landed weight at $1.51 landed price per pound.

on coho rather than chinook would be hurt more. But although most commercially caught chinook seem to be taken by the most productive fishers it is not correct to conclude that they would not suffer some losses. Preliminary examination of the species composition of salmon catch by commercial vessel indicates that many highly productive chinook fishers also catch large amounts of coho.

Other potential losers are ports that are heavily oriented toward the commercial fishery and lack the facilities to support tourism. The timing of seasons may favor some geographical areas over others. The recreational bag limits may promote one recreational mode (charter or private), but not the other.

It does appear, however, that a modest reallocation of the allowable coho catch from commercial to recreational fishers could improve coastal community income, provided wise choices are made about the accompanying regulations and coho stock sizes increase above the depressed levels recently experienced. The determination of what is reasonable will require more detailed consideration of specific alternatives.

Table 22. Total local economic (income) impacts of ocean salmon recreational fishing.

	Astoria	Newport	Coos Bay
	(In $ per angler day)		
Destination expenditures			
Private boats	45.92	45.92	45.92
Charter boats	60.61	60.61	60.61
Impacts on household income			
Private boats	40.41	36.31	38.39
Charter boats	66.69	59.57	62.88
Average	46.08	41.47	43.77

Note: 78% private boats, 22% charter boats.

LITERATURE CITED

Brown, W.G., C. Sorhus and K.C. Gibbs. Estimated expenditures by sport anglers and net economic values of salmon and steelhead for specified fisheries in the Pacific Northwest. Oregon State University, Corvallis. 1980.

Coppedge, R.O. and R.C. Youmans. Income multipliers in economic impact analysis—myths and truths. Special Report 294. Cooperative Extension Service. Oregon State University, Corvallis. June, 1970.

Crutchfield, J.A. and K. Schelle. An economic analysis of Washington ocean recreational salmon fishing with particular emphasis on the role played by the charter vessel industry. University of Washington, Seattle. For the Pacific Fishery Management Council. January, 1979.

King, D.M. and K.L. Shellhammer. The California interindustry fisheries (CIF) model: An input-output analysis of California fisheries and seafood industries, Volume II. California Sea Grant College Program Working Paper No. P-T-6. Center for Marine Studies. San Diego State University. November, 1981.

Lewis, E. Economic multipliers: Can a rural community use them? WREP 24. Western Rural Development Center. Oregon State University, Corvallis. 1979.

Mandelbaum, T.B., S.B. Wood and B.A. Weber. Sectoral output multipliers for rural counties—lessons from Oregon's input-output studies. EC 1166. Oregon State Extension Service. Oregon State University, Corvallis. February, 1984.

Meyer, P.A. Net economic values for salmon and steelhead from the Columbia River system. NOAA Technical Memorandum NMFS, F/NWR-3. National Marine Fisheries Service. Portland, Oregon. June, 1982.

Oregon Department of Fish and Wildlife. Coho plan status report. February 1, 1985a.

Oregon Department of Fish and Wildlife. Progress report on the economic aspects of the recreational/commercial allocation of coho salmon in the ocean fisheries. August, 1985b.

Pacific Fishery Management Council. Framework amendment for managing the ocean salmon fisheries off the coasts of Washington, Oregon, and California commencing in 1985. Portland, Oregon. June, 1984.

Radtke, H., S. Detering and R. Brokken. A comparison of economic impact estimates for changes in the federal grazing fee: Secondary vs. primary data I/O models. *Western Journal of Agricultural Economics*, 10(2). December, 1985.

Rettig, R.B. A Comprehensive Study of the ocean salmon fisheries off the coasts of California, Oregon, and Washington and related inside impacts. Oregon State University, Corvallis. For the Pacific Fishery Management Council. April, 1984.

Rettig, R.B. and B.A. McCarl. Potential and actual benefits from commercial fishing activities in making economic information more useful for salmon and steelhead production decisions. NOAA Technical Memorandum NMFS F/NWR-8. National Marine Fisheries Service. Portland, Oregon. July, 1984.

Siverts, E., C. Palmer, and K. Walters. IMPLAN users' guide. U.S. Forest Service. Fort Collins, Colorado. September, 1983.

Survey Research Center. 1977 Oregon angler survey. Oregon State University. Corvallis, Oregon. For the Oregon Department of Fish and Wildlife. September, 1978.

The Market for Salmon

Richard S. Johnston

*Professor of Agricultural and Resource
  Economics
Oregon State University*

In much of the discussion of salmon markets, emphasis has been on how events in those markets (landings, regulations) have affected variables of interest (prices, costs). The major point of this paper is that trends in individual markets, including salmon markets, are affected not only by conditions in those and related markets, but also by conditions in the total economy. For convenience, economists classify these conditions as *microeconomic* and *macroeconomic* conditions. Besides the need to consider both in conducting thorough economic analysis, an additional reason for distinguishing between the two is that participants in individual markets may be more successful in influencing microeconomic than macroeconomic variables.

Consider price patterns over the past few years. Figure 1 graphs indices of producer prices of coho salmon, livestock, and poultry and eggs, expressed in real terms, between 1964 and 1985. With some exceptions, the patterns are very similar. Consider, for example, the large increase in coho prices that occurred in 1973—larger, in percentage terms, than experienced in the other two markets—and the subsequent dramatic decline after 1979. Attempts to explain these price patterns must consider both

155

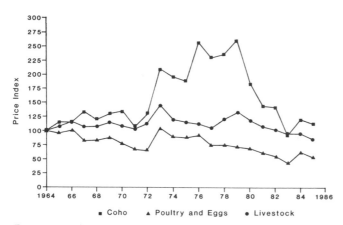

Figure 1. *Indices of real prices received by fishers and farmers, 1964-85.*
* Nominal prices deflated by the GNP implicit price deflator and indexed with 1964 = 100.
Sources: National Marine Fisheries Service, *Fisheries of the United States*, various issues; *U.S. Statistical Abstract*, 1985; Agricultural Statistics Board, National Agricultural Statistics Service, U.S. Department of Agriculture, *Agricultural Prices*, various summary issues.

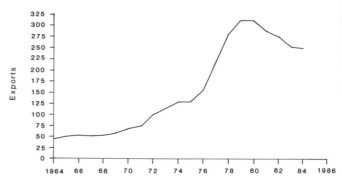

Figure 2. *Index of U.S. exports of seafoods, in real value terms, 3-year moving averages, 1964-84.*
* Nominal values deflated by the GNP implicit price deflator and indexed with 1972 = 100.
Sources: National Marine Fisheries Service, *Fisheries of the United States*, annual volumes; U.S. Government Printing Office, *Economic Report of the President, 1986.*

macroeconomic factors (probably accounting for changes in salmon prices that are similar to price changes elsewhere in the economy) and microeconomic factors (probably accounting for deviations from patterns elsewhere in the economy). Both factors are also crucial to understanding current conditions in salmon markets.

MACROECONOMIC CONDITIONS AND U.S. SEAFOOD TRADE

The need to go beyond microeconomic forces to understand trends in particular markets can be illustrated by consideration of events in the U.S. fisheries sector overall. Specifically, consider the effects of extending jurisdiction over fishery resources through the 1976 Fishery Conservation and Management Act (FCMA) on U.S. trade in seafoods. Prior to implementation of the Act it was expected by some that the U.S., long a net importer of seafoods, would experience increased harvests and, thus, both increased exports and reduced dependence on imports. What actually happened? Between the 3-year period just prior to the FCMA and the 1980-82 period, the average annual harvest by U.S. commercial fishers rose by more than 30 percent while the real value of U.S. seafood exports increased by over 120 percent. This suggests that the FCMA did, in fact, have the expected

impacts on landings and exports. An inspection of U.S. seafood exports (3-year moving averages, to smooth out annual fluctuations) over the 1964-84 period (Figure 2) appears to confirm the view that the FCMA led to increased seafood exports. Between the mid-1960s and mid-1970s, seafood exports grew gradually and then leveled off. Following passage of the FCMA, exports increased dramatically. They have declined in recent years but are still substantially above their mid-1970s level. It would appear that the FCMA has had its expected impact. However, an examination of data in other sectors of the U.S. economy suggests that caution must be used in attributing changes in trade activity to extended fisheries jurisdiction. Figures 3 and 4 depict U.S. exports (in real value terms) of agricultural products and all merchandise, respectively, for the 1964-84 period. Comparison of the trends in these sectors with those in the fisheries sector reveals almost identical patterns. It would be difficult to attribute increases in exports of agricultural products and of all merchandise, taken collectively, to the FCMA; thus the case for a relationship between the FCMA and U.S. seafood exports is weakened.

Imports, too, have followed similar patterns, even in the seafood sector (Figures 5, 6, and 7). Indeed, with respect to both imports and exports, the seafood sector looks very similar to other sectors of the U.S. economy, as measured by changes in economic activity. Here, then, is a case where conditions in individual sectors of the economy, including the seafood sector, reflect changes in over-

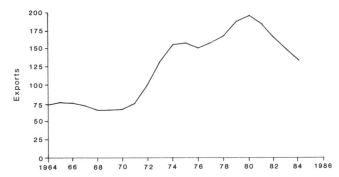

Figure 3. Index of U.S. exports of agricultural commodities, in real terms, 3-year moving averages, 1964-84.
* Nominal values deflated by the GNP implicit price deflator and indexed with 1972 = 100.
Source: U.S. Government Printing Office, *Economic Report of the President*, various issues.

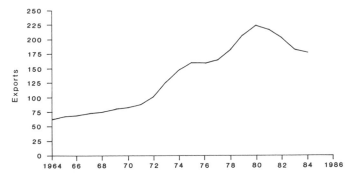

Figure 4. Index of real U.S. international transactions, merchandise exports (excluding military), 3-year moving average, 1964-84.
* Nominal values deflated by the GNP implicit price deflator and indexed with 1972 = 100.
Source: U.S. Government Printing Office, *Economic Report of the President*, various issues.

all, macroeconomic conditions. During the 1960s the U.S. economy experienced a period of uninterrupted expansion, with increases in gross national product and decreases in unemployment. During the mid-1970s the U.S. experienced its most severe recession since the 1930s. Recovery began in 1976, the year of passage of the FCMA. This was followed during the 1980-82 period by another slump and subsequent recovery. It is unlikely that seafood markets are immune from such changes in overall economic conditions. While the FCMA may have played a role in U.S. seafood trade, macroeconomic factors have undoubtedly been extremely important (see Johnston and Siaway 1985).

SALMON MARKETS: MICROECONOMIC AND MACROECONOMIC FACTORS

Now consider developments in salmon markets, including global harvests of salmon. Figures 8a and 8b, taken from Yamamoto (1986), depict landings of the four major salmon producers: Japan, the U.S., the U.S.S.R., and Canada. Note particularly the relationship between U.S. and Japanese harvests since 1975.

The FCMA strengthened earlier U.S. legislation designed to control interception of North American salmon by the Japanese distant water fleet. This, together with management of the domestic fishery, enhancement programs, and favorable

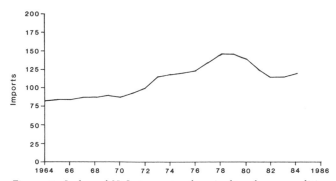

Figure 5. Index of U.S. imports of agricultural commodities, in real terms, 3-year moving averages, 1964-84.
* Nominal values deflated by the GNP implicit price deflator and indexed with 1972 = 100.
Source: U.S. Government Printing Office, *Economic Report of the President*, various issues.

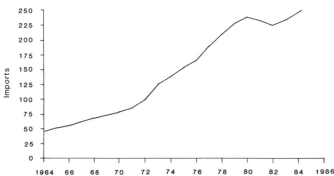

Figure 6. Index of real U.S. international transactions, merchandise imports (excluding military), 3-year moving averages, 1964-84.
* Nominal values deflated by the GNP implicit price deflator and indexed with 1972 = 100.
Source: U.S. Government Printing Office, *Economic Report of the President*, various issues.

157

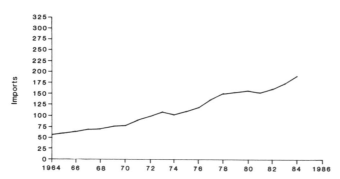

Figure 7. Index of U.S. imports of seafoods, in real value terms, 3-year moving averages, 1964-84.

* Nominal values deflated by the GNP implicit price deflator and indexed with 1972 = 100.

Sources: National Marine Fisheries Service, *Fisheries of the United States*, annual volumes; U.S. Government Printing Office, *Economic Report of the President*, various issues.

climatic and oceanographic conditions, led to large runs of salmon being made available to the Alaska fishery. At the same time, the U.S.S.R. reduced its allocation of salmon to the Japanese fleet. The data reveal a drop in the Japanese harvest between 1975 and 1978, and an increase in both Canadian and

U.S. salmon landings. Two consequences of this are relevant to the current discussion: an increase in Japan's imports of salmon and a decrease in Japan's exports of canned salmon. Consider each of these in turn.

Figure 9 depicts U.S. exports of fresh and frozen salmon to Japan during the 1964-85 period. The U.S. presence in the Japanese market increased dramatically following the decline in the Japanese harvest. The U.S. is now the major supplier of Japan's salmon imports, accounting for 87 percent of the total in 1985 (FAO/GLOBEFISH 1986). Furthermore, Japan itself accounts for over half of the total fresh and frozen salmon imported internationally (Table 1) and, in addition, imports 10,000 mt of salted salmon roe annually (FAO/GLOBEFISH 1986). A number of factors lie behind this trade activity but an important one is the change in nationality of salmon harvesters in the northeast Pacific Ocean. With increased landings by U.S. fishers and decreased landings by Japanese fishers, this trade activity is not surprising, and is intensified by reductions in fishing rights in U.S.S.R. waters. However, with increased catches of fall

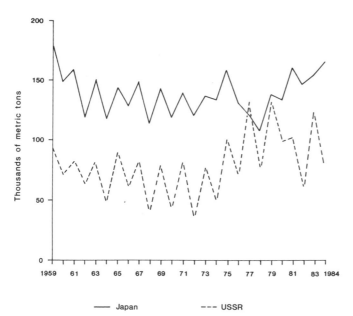

— Japan   --- USSR

Figure 8a. Total salmon catch taken in the northwest Pacific.

Data source: United Nations Food and Agriculture Organization *Yearbook of Fishery Statistics.*

From: Tadashi Yamamoto, *Overview of Salmon Production in the World with Particular Reference to Pacific Salmon.* College of Economics, Nihon University, Tokyo.

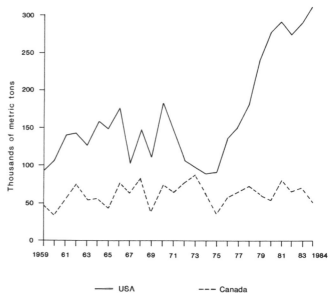

— USA   --- Canada

Figure 8b. Total salmon catch taken in the northeast Pacific.

Data source: United Nations Food and Agriculture Organization *Yearbook of Fishery Statistics.*

From: Tadashi Yamamoto, *Overview of Salmon Production in the World with Particular Reference to Pacific Salmon.* College of Economics, Nihon University, Tokyo.

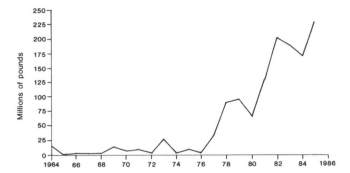

Figure 9. U.S. exports of fresh and frozen salmon to Japan, 1964-85.

Source: National Marine Fisheries Service, *Fisheries of the United States*, various issues.

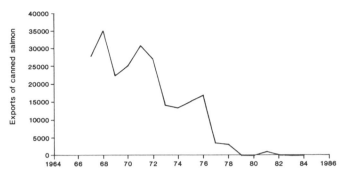

Figure 10. Real value of Japan's exports of canned salmon, 1967-83.

* Million yen, deflated by Japan's wholesale price index, and indexed with 1975 = 100.

Sources: Ministry of Agriculture, Forestry and Fisheries, *Fisheries Statistics of Japan*, various issues; Organization for Economic Co-operation and Development, *Review of Fisheries in OECD Member Countries*, various issues.

chum salmon off Hokkaido, in response to the hatchery program there (Table 2 and Figure 8b), Japan's dependence on imports may decline. Nonetheless, here is an example of an event *within* the salmon sector (a substitution of U.S. for Japanese fishers in the northeast Pacific) affecting trade patterns.

A second effect, as indicated above, is the drop in importance of Japan as a global supplier of canned salmon. Figure 10 reveals that Japanese exports of canned salmon were declining prior to the mid-1970s, but events of that period effectively removed Japan from its role as one of the four major suppliers of canned salmon (along with the U.S., Canada, and the U.S.S.R.) in global markets. Between 1961 and 1970 Japan accounted for over 55 percent of canned salmon traded internationally. Since 1979 its role has been insignificant. Whether this changes in the future will depend, in part, on what happens to Japanese landings. In any

event, here is another phenomenon the explanation of which lies largely within the salmon sector.

What about trade involving other countries? Figure 11 depicts U.S. exports of fresh and frozen salmon to France, another important importer of the U.S. product. Is the pattern revealed there the result of events within the salmon sector? No doubt, in part. However, consider the role of a factor outside of that sector: exchange rates. As suggested by Figure 12, there is a relationship between salmon imports and exchange rates. This is reasonable: the more "expensive" U.S. currency, the more expensive are goods for which payment must be made in U.S. funds. Of course other factors are important, but it does appear that macroeconomic variables do play a role in the seafood sector. Indeed it is likely that the appearance in the U.S. market of imported, farm-raised Atlantic salmon from Norway in the early 1980s is not independent of the strength of the U.S. dollar at that time.

Thus, macroeconomic variables are important to the salmon market. Conditions in other markets are also important. This can be illustrated by re-considering the Japanese case. Figure 13 portrays an increase in the consumption of salted salmon, on a per-household basis, in Japan over the 1964-84 period. A detailed econometric analysis (currently being conducted at Oregon State University) should help identify reasons for this. However, other consumption trends may also provide some insights. An inspection of data on pork consumption (Figure 14), for example, reveals increases similar to

Table 1. Fresh and frozen salmon imports
(In 1,000 mt)

	1981	1982	1983	1984	1985
Japan	72	108	99	93	116
France	17	19	22	24	23
Canada	12	6	6	10	15
USA	3	5	7	10	12
UK	6	6	7	7	7
Germany FR	4	4	5	6	6
Sweden	6	7	7	7	7
Denmark	5	6	6	6	6
Others	12	12	15	12	12
Total	137	173	173	175	204

Source: FAO/GLOBEFISH.

Table 2. Japan's chum salmon catches off Hokkaido.

Year	Number of Fish
1978	13,146,600
1979	18,903,000
1980	15,443,900
1981	21,926,000
1982	20,039,776
1983	22,497,832
1984	19,278,693[a]
1985	29,497,712[a]

Source: Statistics Bureau, Management and Coordination Agency, Japan.

[a] To November 20.

those for salted salmon. Beef and poultry consumption has also risen. Increases in the consumption of these meat items has accompanied increases in real income levels in Japan and government programs to encourage production of these goods. That is, both supply and demand factors probably lie behind these trends. It should be noted that these data are for consumption in the home and exclude consumption in restaurants.

With increases in the consumption of these foods, where have decreases occurred? Two foods whose per-household consumption has declined over the same period are whale meat and rice (Figures 15 and 16). These have been selected for discussion here for two reasons: *(1)* to suggest possible interdependencies, and *(2)* to suggest the need to exercise caution in imputing cause and effect relationships to variables which appear to have similar temporal patterns.

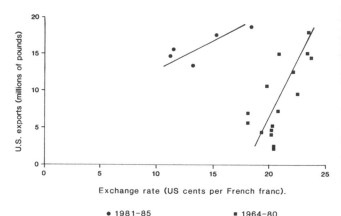

● 1981-85   ■ 1964-80

Figure 12. U.S. exports of fresh and frozen salmon to France and the dollar/franc exchange rate.
Sources: National Marine Fisheries Service, *Fisheries of the United States*, various issues; U.S. Government Printing Office, *Economic Report of the President*, various issues; *U.N. Statistical Yearbook*, various issues.

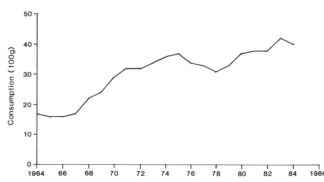

Figure 13. Annual consumption of salted salmon per household in Japan, 1964-84.
Source: Statistics Bureau, Management and Coordination Agency, Japan, *Annual Report on Family Income and Expenditure Survey*, 1984.

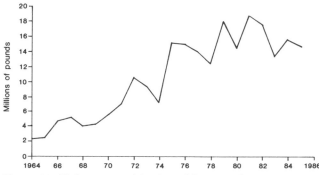

Figure 11. U.S exports of fresh and frozen salmon to France, 1964-85.
Source: National Marine Fisheries Service, *Fisheries of the United States*, various issues.

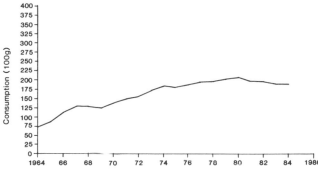

Figure 14. Annual consumption of pork per household in Japan, 1964-84.
Source: Statistics Bureau, Management and Coordination Agency, Japan, *Annual Report on Family Income and Expenditure Survey*, 1984.

One might hypothesize that, with rising incomes, many Japanese consumers have switched away from the traditional dishes, of which rice is a part, to more western dishes. The quantity of rice-based dishes demanded has declined as consumers substituted away from them in favor of dishes containing larger portions of meat, poultry, and seafoods. This could explain decreasing consumption of rice while salmon consumption was rising. Is this also what happened in the case of whale meat? That is, did rising incomes lead to a switch away from whale meat and toward salmon? Perhaps, but a more likely explanation is that the data on whale meat consumption reflect reductions in supply resulting from international restrictions on whaling.

Thus, while there may be interdependencies among markets for foods in Japan, it is important not to confuse cause and effect relationships. Trends in rice and whale meat consumption have been similar, but they are probably the *result* of activities in the salmon market in the case of rice and a *force behind* salmon consumption in the case of whale meat. However, the central point here is that, while macroeconomic factors are important in salmon markets, as are factors within the salmon sector, interdependencies among markets also play a role.

MARKET RELATIONSHIPS BETWEEN PACIFIC AND ATLANTIC SALMON

Now consider the role played by events within salmon markets themselves. There is interest in the effect of farm-raised Atlantic salmon on markets that have been dominated by Pacific salmon. This issue is addressed in a recent University of Alaska Sea Grant report (Rogness and Lin 1986). Researchers surveyed a sample of sixty salmon wholesalers across the U.S. that purchase both species, in order to obtain information on *(1)* their perceptions of the relationship between the Pacific and Atlantic species; and *(2)* their marketing activities. Tables 3-6 and Figure 17 show the responses to some of the questions asked. The research findings suggest that over 40 percent of the Atlantic salmon sales of wholesalers interviewed are made to restaurants (Table 3). This is an interesting—although not surprising—finding and takes on particular significance

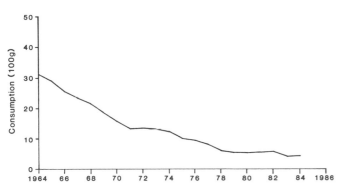

Figure 15. *Annual consumption of whale meat per household in Japan, 1964-84.*
Source: Statistics Bureau, Management and Coordination Agency, Japan, *Annual Report on Family Income and Expenditure Survey*, 1984.

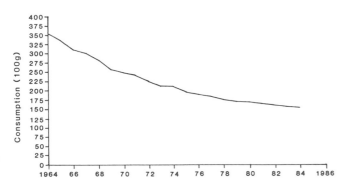

Figure 16. *Annual consumption of nonglutinous rice per household in Japan, 1964-84.*
Source: Statistics Bureau, Management and Coordination Agency, Japan, *Annual Report on Family Income and Expenditure Survey*, 1984.

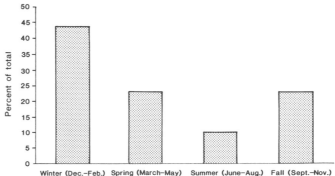

Figure 17. *Average composition of total annual sales (at wholesale) of fresh pen-raised Atlantic salmon by seasonal percent of total (survey results).*
Source: Rogness and Lin 1986.

161

when it is recognized that the growth of the restaurant industry, though positive, is lower than it was 10 years ago. Figure 17 suggests that Atlantic salmon appear more prominently in the market during those seasons when the presence of fresh Pacific salmon is lowest. However, the findings reported in Table 4 indicate that some buyers may believe that there will be little substitution between the Atlantic and Pacific species while others perceive that the two would compete with each other if they appeared at the same time and, thus, this latter group purchases the two species during different months of the year.

This is only one interpretation of these findings, however. To address the substitutional issue more directly, the researchers asked the respondents to identify perceived advantages of Pacific over Atlantic species and vice versa. Tables 5 and 6 report the results. Notice that 24 of the 60 respondents indicated that price favors the Pacific species. This suggests an alternative interpretation of the Table 4 finding that 18 of 52 firms indicated that they purchase fresh Atlantic salmon when fresh Pacific salmon is not available. If the Atlantic and Pacific species are close substitutes in the fresh salmon market, then having fresh Atlantic salmon available when fresh Pacific salmon is not available may effectively increase the demand for the Pacific species. This would be the case if having a year-round, regular supply is important to the retailer, which is very likely the case for the restaurant industry. This suggests another possible interpretation of the finding that 25 of the respondents' purchases of Atlantic salmon are independent of the availability of fresh Pacific salmon. Rather than being perceived as independent in demand, these respondents may see the two species as substitutes in the marketplace and, thus, purchase according to price and availability.

In any event, important changes are occurring in salmon markets. These changes, as argued here, have both macroeconomic and microeconomic dimensions.

Table 4. *Relationship between wholesale purchases of fresh pen-raised Atlantic salmon and availability of Pacific salmon.*

When Purchased	Responses
Independent of the availability of fresh Pacific salmon	25
When fresh Pacific salmon is not available	18
In place of Pacific salmon for all markets	4
In place of Pacific salmon for the following markets:	5
Restaurants	4
Retail	4
Seafood markets	4
Other wholesalers	4

Source: Rogness and Lin 1986.
Note: Data from survey results.

Table 5. *Perceived advantages of fresh wild Pacific salmon over fresh pen-raised Atlantic salmon.*

Advantage	Responses
Price	24
Flesh characteristics (color, firmness, less fat)	19
No advantage	8
Large fish	5
Volume	2
Ease of ordering	1
Never had any fresh Pacific salmon	1

Source: Rogness and Lin 1986.
Note: Data from survey results.

Table 3. *Average composition of annual wholesale sales of fresh pen-raised Atlantic salmon by category of retailers.*

	Percentage of Total Annual Sales
Restaurants	42.3
Other Wholesalers	28.8
Retail Seafood Markets	19.1
Supermarkets	5.5
Others	4.7

Source: Rogness and Lin 1986.
Note: Data from survey results.

Table 6. *Perceived advantages of fresh pen-raised Atlantic salmon over fresh wild Pacific salmon.*

Advantage	Responses
Year-round availability	28
Freshness	26
Consistency	19
Size	12
Price	11
Longer shelf life	5
None	4

Source: Rogness and Lin 1986.
Note: Data from survey results.

ACKNOWLEDGMENTS

This work is a result of research sponsored, in part, by the Oregon State University Sea Grant College Program supported by NOAA, Office of Sea Grant, U.S. Department of Commerce, under Grant Number NA85AA-D-SG065. The author is grateful to Tadashi Yamamoto and R. Bruce Rettig for valuable comments on an earlier draft of this paper.

LITERATURE CITED

FAO/GLOBEFISH Highlights. 1986. A quarterly update based on the GLOBEFISH databank (3/86: 40 p.).

Johnston, R.S. and A. Siaway. 1985. Extended fisheries jurisdiction and seafood trade, Proceedings of the second conferencc of the International Institute of Fisheries Economics and Trade, Volume 2, Christchurch, New Zealand, August 20-23, 1984, Oregon State University Sea Grant College Program Report ORESU-W-84-001, pp. 17-29.

Rogness, R.V. and Biing-Hwan Lin. 1986. The marketing relationship between Pacific and pen-raised salmon: A survey of U.S. seafood wholesalers. University of Alaska. Alaska Sea Grant Report 86-3.

Yamamoto, Tadashi. 1986. Personal correspondence.

Sport Fishing Economics Today and Tomorrow

David B. Rockland

Director of Economics
Sport Fishing Institute
Washington, D.C.

INTRODUCTION

The purpose of this paper is to discuss the key economic concepts pertaining to sport fishing and their relevance to management decisions. A little over a year ago I had the pleasure of taking a unique job—an economic advocate for sport fishing. In the course of the year I have seen how important economics currently is and how important it can be to the management of natural resources. Managers are asking economic questions in their decision-making process.

Many people question the usefulness of economics and economists. We are the subject of a great many jokes. An advisor to a recent U.S. president said, "An economist is someone who, upon seeing something work in practice, wonders if it will work in theory." Economists are also noted for their lack of agreement. "If you took all the economists in the world and laid them end to end, they would all point in different directions." The morbid version of that joke is: "If you took all the economists in the world and laid them end to end, it would be a good thing." People even make jokes about people who listen to economists. For example, "It is interesting that the same people who dismiss gypsy fortunetellers are the first to believe economists."

From a serious perspective, sport fishing economics is a term that troubles some people. The terms "sport fishing" and "economics" are felt not to go together. Sport fishing is just fun or frivolous, and has no economic value. The major point of this paper is that sport fishing is of tremendous economic significance and the economic benefits of sport fishing must be considered in the management of all fishery resources, including salmon.

WHAT IS ECONOMICS?

Any student in an economics course hears the same thing the first day of class: "What is economics?" The general answer is that economics is the study of the allocation of scarce resources among competing wants. Managers can use economics as a tool to help them make decisions concerning the resources they manage. There are two general economic questions relevant to managers: allocation of natural resources—in this case, fish—and allocation of financial resources—the investment of money.

The allocation of fish is a question of which user group can use the resource. Should commercial fishing be prohibited in favor of sport fishing? Do the fish have a higher value to the economy or society as sport fish or commercial fish? These are very current and contentious questions regarding salmon as well as other species. The allocation of fish is an issue of most importance in the marine or saltwater environment. It is interesting to point out that managers long ago made the intuitive decision that the American people were, for the most part, better off if inland fisheries were allocated to sport fishing. Managers made similar decisions regarding wildlife. We no longer have market hunting for deer or waterfowl. Ostensibly, these managers made resource allocation decisions without the benefit of economists.

The allocation of financial resources is a question of how to spend one's budget. For example, should I stock this stream or that stream, this type of salmon or that type of salmon? Should trout fishing be delayed one month to allow for the safe passage of salmon smolts? Are the economic costs of losing a month of trout fishing offset by the long-term gains of a higher survival rate of salmon smolts?

Allocation of fish and money are issues that the economist can address. The derivation of the answers is what we call sport fishing economics. A type of question that is often confused with economic questions is the issue of equity or fairness. For example, an economist can tell a decision maker whether the economic benefits from the allocation of salmon to Native fishing exceed the economic costs to sport fishing, but not whether such an allocation is fair or equitable given the history of the rights of American Natives on the west coast and the needs of sport fishing and the sport fishing industry.

WHAT ARE ECONOMIC BENEFITS?

To undertake the analysis of an allocation question requires an understanding of the economic benefits and costs. A manager might ask, "What are the economic benefits to sport fishing from the decommercialization of salmon?" The term "economic benefits" includes a number of concepts.

Two basic benefits result from the fishery resource—economic value and economic impact. Economic value is the answer to the question: "How much value do people place on the resource?" An economic impact is the answer to the question: "What is the economic activity generated by the use of the resource?" While these two benefits are distinct, they are not entirely inseparable. Neither has greater merit as an economic concept than the other; rather, they address two distinct questions.

Each question is important for different reasons. Local and state government officials often base decisions on the economic impact in their region which translates into jobs, income, and tax receipts. Economic value, on the other hand, is the value people place on the resource. This concept is certainly important to those who value the fishing experience and the fish stock as a component of that experience. In addition, it has become a concept with increasingly greater application in fisheries management. However, economic value remains a vague concept to many people regardless of its definite existence.

Economic value

Values attributable to the fishery resource exist for both users and nonusers of the resource. User benefits take three forms—consumptive use, nonconsumptive use, and indirect use—while nonuser benefits are defined as option or existence benefits.

Consumptive use. Consumptive use benefits are enjoyed by fishers and other "consumers" of the fishery resource. ("Catch-and-release" fishing is considered a consumptive use because the actual activity the user undertakes is most similar to "catch-and-keep" fishing.)

The fact that anglers go fishing in preference to another activity, and that they spend money, time, and effort doing so, indicates that they value the opportunity to go fishing and the fish stock. In addition, many would be willing to pay additional money rather than be excluded from using the resource; this is a further indication of consumptive use benefit. The entire willingness to pay is the total value to the resource user.

Nonconsumptive use. Just as anglers place a value on fish and fishing, so do nonconsumptive users such as photographers, snorkelers, aquarium visitors, and others who receive value from directly viewing the resource. While wildlife may more commonly offer nonconsumptive user benefits such as bird-watching, the popularity of nonconsumptive snorkeling and scuba diving attests to the validity of the concept of nonconsumptive user benefits for fisheries, as do the people who watch salmon on their annual spawning run. Like consumptive-use benefit, benefit to nonconsumptive users is indicated by a willingness to pay before foregoing the activity, as well as by the expenditures, time, and effort made to undertake the activity.

Indirect use. An indirect use is when people do not come into contact with the resource, but still derive personal satisfaction from it. Indirect use of the fishery resource includes reading about fish, viewing pictures of fish, watching television specials about fish, and related activities. People spend money (on, for instance, books, magazines, television) to use the resource indirectly. This in itself is an expression of value, and there may be additional monies that an indirect user would be willing to pay before having to forego the opportunity for indirect use.

Nonuse benefits are also referred to as intrinsic values and result from sentiments about the resource that are not related to current use. Nonuse benefits are categorized as option value and existence value. Option value reflects uncertainty about future resource use and represents the value the individual places on the availability of resource use in the future. In other words, individuals value the

option of having the resource available in the future in the event they want to use it.

Some people may value a fishery resource even if they know they will never use it themselves. This is sometimes called existence value. Existence value is generally motivated by altruism or the unselfish concern for other people or the fishery resource. One form of existence value is bequest value, which captures the desire to endow the resource to future generations. Other forms may be motivated by altruism toward the fishery resource itself, in the same way that people place a value on knowing that whales and bald eagles exist in the wild and are not extinct.

It may sometimes be difficult to distinguish between indirect use value and existence value. While economics does not offer a clear-cut distinction, indirect use value can be categorized as some form of indirect contact with the resource or indirect personal enjoyment of it. Existence value results only from the knowledge of existence. The line between these two types of benefits is somewhat obscure, although both clearly exist and are distinguishable in most cases.

Economic impact

The expenditures made in the course of consumptive, nonconsumptive, or indirect use may impact the local, state, and/or national economies. These impacts are expressed in terms of jobs, sales, or wages and salaries. Total economic impact is based on expenditures made by consumptive, nonconsumptive, and indirect users, but exceeds these expenditures since sales in one industry impact not only that industry but also industries that supply goods and services to it.

Three levels of economic impact are discernible:

Direct impact. The initial purchase by the recreational angler; for example, the purchase of a fishing rod.

Indirect impact. The purchase of inputs by the directly impacted business to produce the goods and services purchased by recreational fishers; in our example, these would include the purchase of graphite, paint, and guides to make the rod. The initial round of indirect purchases has further indirect impacts as the suppliers to the first businesses make purchases to meet that demand; in our example, further indirect impacts might include the purchase of pigments by the paint supplier and aluminum by the guide supplier.

Induced impact. The purchase of goods and services resulting from the wages paid by the directly and indirectly impacted businesses, such as purchases by households with wages made by employees of the tackle, graphite, paint, guides, pigment, and aluminum manufacturers. Note that induced impacts have, in turn, additional indirect and induced impacts.

Economic impact is a very important concept to fisheries managers. It measures the activity in the region's economy associated with the use of the resource, which is important to the economic vitality of the region. However, economic impact cannot be considered a true benefit of the fishery resource because economic impact results from the costs incurred to use the resource. A true economic benefit is the value of a good or service less all the costs of creating it. Economic impact is not therefore a true economic benefit, because it results from the cost of creating the product (in this case, a fishing trip). However, economic impact is the measure of the benefit that industry and wage earners receive from the use of the fishery resource. Fishery management decisions should be considered in terms of industry impact as well as user and nonuser values because industry impact translates into jobs, income, and sales.

A key point in understanding these various concepts is that they are not all additive. One does not add the direct, indirect, and induced impacts to the various types of user and nonuser values to derive total benefits. Rather, the sum of the direct, indirect, and induced impacts is the total economic activity of sport fishing and responds to the questions: "How is the economy affected? How many jobs are affected?" One user benefit is how much the sport fishers value the opportunity to go sport fishing. The other benefits reflect amounts nonusers value the resource either in terms of an option to use it some day, for their offspring to use it, or for other reasons.

If economics is to be properly considered in the management process, all of these economic concepts need to be addressed. In that way all the possible benefits and costs of an action are understood. Failure to consider all of these aspects will result in the possibility of erroneous choices in the management process, from the perspective of economics.

WHAT ORDER OF MAGNITUDE ARE THE ECONOMIC BENEFITS OF SPORT FISHING?

I work for the sport fishing industry. That industry comprises businesses that supply goods and services to sport fishers. It includes hotels, motels, restaurants, tackle manufacturers, boat manufacturers, magazine publishers, and gasoline distributorships. The industry was worth about $28.2 billion at retail in 1985. In other words, on the order of $28.2 billion was spent for sport fishing. In terms of the three types of economic impact I described, the direct impact of sport fishing on the national economy is $28.2 billion. If the indirect and induced impacts are added, the total economic impact of sport fishing on the national economy is between fifty billion and seventy-five billion dollars. Expressing those same numbers in terms of employment, on the order of 800,000 jobs exist in the United States as a result of sport fishing.

Sport fishing is presently the number two form of recreation among adult Americans. The value placed on sport fishing by sport fishers is enormous. For freshwater fishing, total value in excess of expenditures has been estimated to be between $23.8 billion and $54.1 billion, with an average of $38.9 billion. In other words, in addition to the amount freshwater anglers spent to go fishing in 1985, they valued the opportunity to do so by an additional $38.9 billion. Of this amount, approximately 28 percent was attributed to trout and salmon fishing, 60 percent was attributed to bass fishing, and the remaining 12 percent was attributed to other sport fisheries.

ARE THE ECONOMIC IMPACTS OF SPORT FISHING IMPORTANT?

The argument is advanced at times that the economic impact of sport fishing should be unimportant in the management of the resource. The reason given is that if sport fishing were not to occur those expenditures made for sport fishing would likely be made on other recreation activities such as hunting, bowling, or movie attendance, and hence the net economic change would be zero. The substitution to other like goods or services is valid for any industry. For example, if commercial fishing were to disappear, people would substitute other food products. Therefore, while employment would decrease for commercial fishers, it would increase for chicken farmers.

If we take this line of argument to its logical end, the net economic impact of any industry at the national level is zero, because consumers who are deprived of that industry's good or service will spend their money on something else. However, our society spends a great deal of effort making sure industries are maintained. We recognize that there are true dislocations to people when industries disappear. Those dislocations occur across regions and across skill types. For example, the loss of sport fishing in one region may be offset with increased sport fishing in another region. However, employees in the sport fishing industry where the loss has occurred are not able to move at no cost to the other region. A loss of sport fishing may be followed by an increase in movie attendance. However, bait and tackle shops cannot instantly become movie theaters, in the same way that commercial fishers cannot instantly become chicken farmers.

There are costs to the loss of an industry, and the sport fishing industry is a very large industry that employs a great many people. Management decisions for the public salmon resource that maintain a healthy sport fishery will maintain a large and viable industry. Conversely, decisions that are made without fully addressing the economic benefits and costs relevant to sport fishing can put a great many people out of work.

CONCLUSION

In conclusion, a wide array of economic benefits result from sport fishing and the fishery resource. These need to be considered as much as possible in the management process. Failure to do so can result in significant losses to users and nonusers of the resource and have substantial industry effects. Proper consideration of all of the economic benefits of the resource can result in the maintenance and expansion of the significant economic benefits of sport fishing. This means keeping an industry with $28.2 billion in retail sales and 800,000 jobs healthy. It also means maintaining and enhancing the values received by users and nonusers of the resource which are likely to be in excess of fifty billion dollars.

Potential Hazard for Spread of Infectious Disease by Transplantation of Fish

John S. Rohovec

Associate Professor of Microbiology
Oregon State University

James R. Winton

Research Microbiologist
U.S. Fish and Wildlife Service
Seattle, Washington

John L. Fryer

Professor of Microbiology & Department Head
Oregon State University

Disease in fish has been recognized as a potentially limiting factor in any fishery enterprise. The diseases receiving greatest attention and of most concern are those caused by infectious agents or microorganisms which enter the host, multiply, cause pathology, and often result in death. Infectious agents of fish include viruses, bacteria, protozoa, and fungi. Approximately twenty species of bacteria are known to infect salmonid fish (Table 1), and numerous higher parasites have been described from diseased fish. Approximately ten viruses have been isolated and described as pathogens of salmonid fish and about four have, in addition, been observed by electron microscopy only (Table 2). Of the microorganisms that cause disease in fishes, those that infect salmonid fish are probably the most studied and best characterized. Although much is known about salmonid diseases and the agents which cause them, important unanswered questions remain.

The diseases of most concern to salmonid culturists are those for which there is no known therapy, among them those caused by viruses. The major viral pathogens of salmon are infectious hematopoietic necrosis virus (IHNV), infectious pancreatic necrosis virus (IPNV), viral hemorrhagic

171

Table 1. *Bacterial pathogens of salmonids.*

Bacterial Species	Disease
Gram Negative Pathogens	
Vibrio anguillarum[1,2]	Vibriosis
Vibrio ordalii[1,2]	Vibriosis
Vibrio salmonicida	Vibriosis
Vibrio alginolyticus	Vibriosis
Aeromonas salmonicida[1,2]	Furunculosis
Aeromonas hydrophila[1,3]	Motile Aeromonad Septicemia
Edwardsiella tarda[3]	Edwardsiellosis
Yersinia ruckeri[1]	Enteric Redmouth Disease
Pseudomonas fluorescens[3]	*Pseudomonas* Septicemia
Cytophaga psychrophila[1]	Bacterial Coldwater Disease
Cytophaga spp.[1]	Fin Rot, Bacterial Gill Disease
Flexibacter columnaris[1]	Columnaris
Sporocytophaga sp.	Salt Water Columnaris
Flavobacterium sp.	Bacterial Gill Disease
Gram Positive Pathogens	
Renibacterium salmoninarum[1,2]	Bacterial Kidney Disease
Lactobacillus piscicola	Pseudokidney Disease
Streptoverticillium	Streptomycosis
Clostridium botulinum[3]	Botulism
Acid Fast Pathogens	
Mycobacterium chelonei[3]	Mycobacteriosis
Nocardia asteriodes[3]	Nocardiosis

[1] Species of bacteria considered major pathogens of fin fish.
[2] Species of bacteria considered obligate pathogens of fin fish.
[3] Species of bacteria which are also associated with human disease.

septicemia virus (VHSV) and *Oncorhyncus masou* virus (OMV). Examples of bacterial pathogens which are difficult or impossible to treat are *Renibacterium salmoninarum* (the etiological agent of bacterial kidney disease [BKD] and the drug resistant forms of gram negative bacteria, especially *Aeromonas salmonicida*, the cause of furunculosis. Histozooic protozoans which infect salmonids also have no known treatments. Those of most concern are *Myxosoma cerebralis*, which causes whirling disease, and *Ceratomyxa shasta*. In addition, the uncharacterized agent of proliferative kidney disease (PKD) has been considered a potential hazard to salmonid culture.

Many of these infectious agents are not distributed worldwide but are endemic in certain geographic regions. For example, IHNV occurs widely in the western portion of North America and in specific areas of Asia. Until recently, the virus was not thought to occur in Europe, but the first isolation of IHNV in Europe was reported in 1987 by F. Baudin Lovrencin in the *Bulletin of the European Association of Fish Pathologists* 7(4):104. The

occurrence of IHNV in Europe is almost certainly a consequence of importation if fish or eggs from a region where the virus is endemic. VHSV, on the other hand, has been isolated only in Europe and is not found in Asia or North America. Neither of these viruses occur in all populations of salmonids within their geographic range. *Renibacterium salmoninarum* is more cosmopolitan, causing disease in most areas where salmonids are reared, but again not all stocks are necessarily infected. Although *M. cerebralis* is found both in North America and in Europe, the agent and the disease exist in limited geographic ranges within these continents.

Salmonid stocks have been extensively transplanted to new locations for many years and in certain instances their pathogens and parasites have been transported along with them. This has resulted in the spread of infectious microorganisms into areas where they were not endemic. For example, infectious hematopoietic necrosis is endemic among populations of salmonids in Alaska and the sockeye salmon appears to be the primary host. Prior to 1971 IHNV did not exist in Japan. This virus was introduced to the Chitose Hatchery, Hokkaido, Japan, with a shipment of sockeye salmon eggs from Alaska. IHNV has subsequently spread to other populations of fish and now occurs among cultured and wild populations of salmonids throughout the country. The rapid dissemination of IHNV throughout Japan is believed to have resulted from the indiscriminate movement of fish and fish eggs, which was frequently based on economic considerations rather than fish health management. IHN virus has also been transported in or on contaminated salmonid eggs from hatcheries in the Pacific Northwest to other areas in the U.S. where the virus is not endemic (Colorado, New York, West Virginia, Minnesota). But, because of aggressive management which included destruction of infected stocks and sanitation of hatcheries receiving eggs, the virus has not become established in fish at any of these locations.

Transplantation of salmonids into South America has been accomplished but their introduction into Chile has also resulted in the spread of *R. salmoninarum*. Bacterial kidney disease was observed in the early 1960s (J.W. Wood, personal communication) and has recently been reconfirmed (Sanders and Barros 1986). It can be stated with certainty that BKD was introduced with the import of eggs because *R. salmoninarum* is a pathogen of salmonids alone and these fish are not native to South America.

The protozoan *M. cerebralis* may also have been spread as a result of international trade of salmonid fish (Hoffman 1970). This myxosporean is believed to have been introduced into the United States from Europe and has subsequently had a limited extension in geographic range within this country. Salmonids imported into New Zealand may have carried *M. cerebralis* and as a result whirling disease has become widespread in that country.

These examples show that transplantation of fish and fish eggs can be responsible for the inadvertent introduction of their pathogens. It should be noted, however, that not all fish pathogens are easily spread to new locations. *Ceratomyxa shasta*, a myxosporean found in the Pacific Northwest of the United States and on the west coast of Canada, has not been reported to have been spread outside of this geographic region (Johnson et al. 1979). Disease is caused by *C. shasta* in salmonids in the Sacramento River system in northern California, in some river systems of Oregon and Washington, including the Rogue, Nehalem, Columbia and certain of its tributaries, and in the Fraser River in Canada. Although the known geographic range of the parasite has been extended within the region, it has not been reported outside this area. But *C. shasta* seems to be unique in this regard; other fish pathogens are more easily spread to new geographic locations.

Potential hazards from introduction of fish pathogens to nonendemic areas extend beyond the obvious one of causing mortality and subsequent economic loss. Loss of markets can occur both because of a lack of fish to sell as a result of mortality, and because importation laws may prohibit the movement of diseased or carrier fish. For example, fish may not succumb to *M. cerebralis*, but if they carry this agent they cannot legally be transported within or into many countries for any reason. Populations of fish which have had no previous contact with a pathogen are frequently more susceptible than those which have had a long association with it. Even seemingly innocuous agents may have an adverse effect on populations of fish which have not been previously exposed to the agent. There are many examples where introduced stocks of fish have been highly susceptible to an endemic pathogen which exhibits low virulence for

Table 2. Viruses associated with diseases in salmonids.

I. *Viruses isolated in cell culture and partially characterized:*
   DNA Viruses
   Herpesviruses
   Herpesvirus salmonis
   Oncorhynchus masou virus (OMV)
   Nerka virus of Towada Lake, Akita and Aomori (NeVTA)
   Yamame tumor virus (YTV)
   RNA Viruses
   Birnaviruses
   Infectious pancreatic necrosis virus (IPNV) (serotypes VR-299, Ab and Sp).
   Reoviruses
   Chum salmon reovirus (CSV)
   Coho salmon reovirus (CSR)
   Rhabdoviruses
   Infectious hematopoietic necrosis virus (IHNV)
   Viral hemorrhagic septicemia virus (VHSV)
   Paramyxoviruses
   Chinook salmon paramyxovirus

II. *Viruses observed by electron microscopy but not isolated in cell culture:*
   DNA Viruses
   Iridoviruses
   Erythrocytic necrosis virus (ENV)
   RNA Viruses
   Retroviruses
   Atlantic salmon fibrosarcoma
   Unclassified viruses
   Salmonid anemia virus
   Atlantic salmon papilloma virus

native strains of fish. It has been documented that genetic crosses of resistant fish with susceptible stock result in an F_1 generation which expresses disease susceptibility (Hemmingsen et al. 1986). It is important to note that, although diseased fish might be confined to a culture facility, the water released from the site is a potential source of infection for feral fish and animals at other aquaculture facilities.

Discussion thus far has concerned the introduction of fish pathogens into uncontaminated waters or susceptible fish into regions where one or more pathogens exist. A fisheries management concept that may be inappropriate though it is widely accepted is the introduction of diseased or carrier fish into areas in which the same disease condition already exists. This practice deserves comment. The introduction of infected or carrier animals increases the number of infectious units or pathogens present and may cause a critical number ("infectious dose") to be reached whereby otherwise resistant fish become diseased. Conversely, by continual introduction of healthy stock it is possible to eliminate or reduce the frequency of a disease (Yamamoto and Kilistoff 1979). Although a pathogen may be endemic in a region, the particular strain (or serotype) causing the disease may be different from strains (or serotypes) which might be introduced. Recent unpublished data, for example, indicate that there are strains of IHNV which exhibit differences in host preference or have species-specific virulence.

Numerous state, regional, national, and international organizations have developed disease control policies which regulate the movement of fish and fish eggs (Fryer et al. 1979, Rohovec 1983). All these policies are concerned primarily, if not exclusively, with diseases of salmon. International policies have been formulated by the European Inland Fishery Advisory Commission, the International Council for the Exploration of the Sea, and the Food and Agriculture Organization of the United Nations. Individual countries also have regulations; in the United States, importation of salmonids is regulated by Title 50, and various regional policies within the U.S. have been or are being developed including those of the Great Lakes Fishery Commission, the Colorado River Drainage, and most recently the Pacific Northwest Fish Health Protection Committee. Individual states also have regulations, often more restrictive than the federal regulation. Most of these policies include those infectious agents which are nonendemic and for which there is no effective control. Before fish or fish eggs are imported, a certification of health is required. Even if these regulations were strictly enforced, the risk of importation of disease agents is always possible. Health certification, determined by inspection of fish, does not insure that lots of fish or eggs are free of pathogens, but only implies that a statistical number of the total population has been included in the sample examined by prescribed methodology.

Concern for the spread of fish diseases and the regulations that control them seem, to the authors, to have undergone an evolution. Thirty years ago there was little concern, perhaps resulting from a paucity of knowledge of the epidemiology of salmonid diseases, and hence limited attention was

paid to the movement of fish stocks by fisheries managers. There followed a period of increased interest in the development and use of fish disease control policies to prevent the further spread of infectious diseases. Stocks of fish were destroyed because they carried an untreatable virus, and importation and movement were limited. But these fish disease control policies were often found to be restrictive and in conflict with fishery management concepts and plans for enhancement of the resource, which frequently required movement of fish or fish eggs from one location to another. At times movement of stocks was prevented because of concern with the spread of disease. Recently, there seems to be a relaxation of the previous attitude. This may be occurring for biological, political, and/or economic reasons. There are examples of successful fish culture programs where disease exists and where there are few or no regulations, but all would be more efficient and profitable with effective fish health management.

All transplantations of fish and fish eggs are associated with some risk of spreading infectious disease. Three fish health management policies should be implemented to minimize risk: (1) the disease history of the stock of fish to be transplanted should be determined prior to movement; (2) when it has been established that the stock has an acceptable health record, only eggs from parents which have been tested and found free of disease should be transported; and (3) disinfected eggs should be hatched and reared in quarantine until the juveniles can be certified to be free of disease. These policies should be followed in all situations where fish are transplanted to areas where diseases of interest are not known to occur.

ACKNOWLEDGMENTS
This work was sponsored by Oregon Sea Grant through NOAA Office of Sea Grant, Department of Commerce, under Grant No. NA85AA-D-SG095 (project no. R/FSD-10). Agricultural Experiment Station Technical Paper No. 8044.

LITERATURE CITED

Amos, K., ed. 1985. *Procedures for the Detection and Identification of Certain Fish Pathogens.* 3rd edition. Fish Health Section, American Fisheries Society. Corvallis, Oregon.

Carlisle, J.C., K.A. Schat, and R. Elston. 1979. Infectious hematopoietic necrosis in rainbow trout *Salmo gairdneri* Richardson in a semi-closed system. *J. Fish Diseases* 2:511-517.

Fryer, J.L., J.S. Rohovec, E.F. Pulford, R.E. Olson, D.P. Ransom, J.R. Winton, C.N. Lannan, R.P. Hedrick, and W.J. Groberg. 1979. Proceedings from a conference on disease inspection and certification of fish and fish eggs. Oregon State University, Sea Grant College Program Publication No. ORESU-W-79-001.

Hemmingsen, A.R., R.A. Holt, R.D. Ewing and J.D. McIntyre. 1986. Susceptibility of progeny from crosses among three stocks of coho salmon to infection by *Ceratomyxa shasta. Trans. Am. Fish. Soc.* 115:492-495.

Hoffman, G.L. 1970. Intercontinental and transcontinental dissemination and transfaunation of fish parasites with emphasis on whirling disease (*Myxosoma cerebralis*). Pages 69-81 in D.F. Snieszko, ed., *A Symposium on Diseases of Fishes and Shellfishes.* American Fisheries Society. Washington, D.C. [Special Publication No. 5.]

Holway, J.E. and C.E. Smith. 1973. Infectious hematopoietic necrosis of rainbow trout in Montana: A case report. *J. Wildl. Diseases* 9:287-290.

Johnson, K.A., J.E. Sanders and J.L. Fryer. 1979. *Ceratomyxa shasta* in salmonids. Fish Disease Leaflet 58. United States Department of the Interior. Division of Fishery Research. Washington, D.C. 11 pp.

Plumb, J.A. 1972. A virus-caused epizootic of rainbow trout (*Salmo gairdneri*) in Minnesota. *Trans. Amer. Fish. Soc.* 101:121-123.

Rohovec, J.S. 1983. Development of policies to avoid the introduction of infectious disease among populations of fish and shellfish. Pages 371-373 in P.M. Arana, ed., Proceedings of the International Conference on Marine Resources of the Pacific. Viña del Mar, Chile.

Sanders, J.E. and M.J. Barros R. 1986. Evidence by the fluorescent antibody test for the occurrence of *Renibacterium salmoninarum* among salmonid fish in Chile. *J. Wildl. Diseases* 22:255-257.

Sano, T., T. Nishimura, N. Okamoto, T. Yamazaki, H. Hanedu and Y. Watanabe. 1977. Studies on viral diseases of Japanese fishes. VI. Infectious hematopoietic necrosis (IHN) of salmonids in the mainland of Japan. *J. Tokyo Univ. Fish.* 63:81-85.

Sano, T. 1976. Viral diseases of cultured fishes in Japan. *Fish Pathology* 10:221-226.

Sippel, A.J. 1982. Great Lakes fish disease control. *Fisheries* 7:18-19.

Wolf, K., M.C. Quimby, L.L. Pettijohn, and M.L. Landolt. 1973. Fish viruses: isolation and identification of infectious hematopoietic necrosis in eastern North America. *J. Fish. Res. Board Can.* 30:1625-1627.

Yamamoto, T. and J. Kilistoff. 1979. Infectious pancreatic necrosis virus: quantification of carriers in lake populations during a 6-year period. *J. Fish. Res. Board Can.* 36:562-567.

Directed and "Inadvertent" Genetic Selection in Salmonid Culture

Results and Implications for the Resource, and Regulatory Approaches

William K. Hershberger

Associate Professor
School of Fisheries
University of Washington

INTRODUCTION

Selection is one of the major processes determining the genetic composition of populations of living organisms. In the natural environment, selection is caused by largely unknown or uncharacterized circumstances, but many of the circumstances which cause selection are partly under human control. In production situations, selection is directed and controlled by people in a manner that will yield organisms useful to them. The culture of salmonids spans this range of situations since it involves the maintenance of a valued natural resource, as well as production of a food commodity. This makes it difficult to obtain an understanding of the effects of selection in salmonid culture and assess the value of its use. In this paper salmonid culture will be divided into categories and the prevalent selection processes highlighted and assessed in regard to their anticipated effect on the genetics of the populations involved.

In reviewing the literature on the use and role of genetic selection in salmonid culture, it becomes apparent that these subjects evoke a wide diversity of viewpoints. On the one hand, some reports suggest that genetic selection of any type is the antithesis of good fish culture practices. Alternatively, other authors conclude that it can be used to solve most of the production problems encountered with salmon and trout. A number of factors contribute to this wide divergence of viewpoints. First, a major source of this confusion can be attributed to lack of information on the influence of natural selection in determining the genetic composition of populations. This is not unique to salmonids; for the most part, our knowledge of the genetic effects of selection is derived from the practice of artificial selection. Second, there is an apparent lack of appreciation for the variety of selective influences that are possible or practical with the diversity of salmonid culture practices now available. For example, culture practices with Pacific salmon can range from placing fertilized eggs into a stream-side incubation box in order to augment natural reproduction to a sophisticated hatchery operation where all phases of the life cycle are very tightly controlled to obtain maximum production efficiency. Since genetic selection is based on the survival and reproductive success of individuals with particular characteristics, the magnitude of any selective effects can vary widely based on the control exercised over these two factors. Finally, since most salmonid fisheries are harvesting adult fish on their return to spawn, selective effects can result from this source also. All of these factors, separately or in combination, have the potential to affect genetic change in salmonid populations.

In order to analyze the importance of different selective factors in determining the genetic composition of salmonid populations, it will be necessary to simplify a very complex system. Culture practices can be divided into general categories on the basis of human influence on the salmonid resource. Wilkins (1981), in a review article on the importance of genetics to aquaculture, developed a three-phase approach based on the level of intervention used for propagation of various species: salmonid culture can be easily categorized on this same basis. The first phase proposed was a "precultivation phase." Organisms in this phase are essentially naturally reproducing and experience minimal disturbance from propagation activities. Salmonid populations that are categorized as wild, or those that are enhanced via habitat improvement, streamside incubation facilities, or spawning channels can be included in this phase. The second phase Wilkins identifies is the "cultivation phase." This phase is characterized by the development of broodstocks

to reproduce the population and the capability of raising animals throughout their life cycle under artificial conditions, if desired. Most of the state, federal, and private salmonid hatcheries could be placed in this category, as well as some of the commercial captive trout and salmon culture facilities. The final category Wilkins developed was the "postcultivation phase." The culture techniques employed with the organisms in this phase of propagation are quite sophisticated and, in many ways, can be equated with current agricultural practices.

It is apparent that the types of genetic analyses needed and the opportunities to use different types of genetic manipulation will vary, depending on the phase of cultivation; these topics are covered in Wilkins (1981). In this paper, only the practice of selection and its potential effects on the genetic constitution of populations will be addressed. Additionally, the various sources of selection within each category of salmonid propagation will be assessed in order to determine their relative influence on the genetics of the populations. Finally, some recommendations will be proposed to guide the use of selection in salmonid culture.

SELECTION—DEFINITION AND GENETIC IMPLICATIONS

The process of selection needs to be defined before delving into its effects on the genetic constitution of populations of salmonids, and some of the basic genetic tenets important to selection need to be clarified. First, simply defined, selection means differential reproductive rates; the average number of offspring reaching breeding age is different for different kinds of individuals. This result can be accomplished by a wide variety of mechanisms that modify the reproductive success of organisms, and that can occur throughout the entire life cycle. If selection is carried out by the conscious effort of a breeder, the process is termed "artificial selection." On the other hand, if reproductive success is not determined by human intervention, organisms are subject to "natural selection." In either case, the process by which genetic change can occur is the same.

Genetic change due to selection is estimated in terms of the "fitness" of a genotype. Although fitness can have a variety of applied meanings, such as fitness for long migrations, avoiding capture, or obtaining food, in the genetic sense it is far more restricted and refers only to relative reproductive success. In its simplest form, fitness can be measured by comparing the number of offspring produced by each genotype that survives to reproduce. For example, if fish of genotype *A* produce an average of fifty offspring that reach maturity while those of genotype *a* produce only forty-five in the same environment, the relative reproductive success (fitness) of *a* is only 90 percent that of *A*. With this magnitude of difference, it is easy to predict the elimination of the *a* genotype from a population after a number of generations. Although this example provides an overview of the way selection operates, several qualifications must be kept in mind in relating this to actual occurrences.

First, the traits that are of most interest in regard to selection are generally determined by many genes. Further, the expression of these genes is affected by environmental circumstances, interactions among the individual genes, and interactions between the genes and the environment. The end product of all these factors is the phenotype, i.e., the appearance of an organism. In shorthand terms this relationship is expressed as:

$$P = G + E$$

where
$$P = \text{Phenotype,}$$
$$G = \text{Genotype, and}$$
$$E = \text{Environment.}$$

Since the phenotype is the unit (natural and artificial) selection acts on, some estimate of the genetic determination of a trait must be made before any genetic change can be ascribed to the effects of selection.

Second, selection can operate in many diverse ways throughout the life cycle to optimize the reproductive success (fitness) of organisms. The genetic and environmental permutations that will yield this result are limitless and, consequently, selection is a very dynamic process. In natural situations the phenotypic result of selection is generally a compromise in the expression of all the traits important to the maximization of fitness. Very infrequently do organisms express genotypes for specific traits to the maximum extent possible. Instead, a combination of "normal" traits is exhibited that allow utilization of the environment to which they are exposed. This is contrary, in a general way, to what occurs or is desired when artificial selection is practiced. Artificial selection usually leads to large changes in relatively few traits.

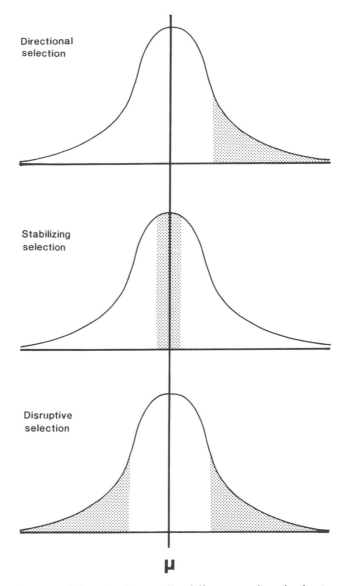

μ

Figure 1. Three fundamentally different modes of selection that occur in populations. The stippled area indicates the phenotypes that are "chosen" for reproduction.

Finally, there are three fundamentally different ways by which selection can operate (Figure 1). Directional selection changes the population in a specific direction and is the type of selection that is most often practiced in stock development programs. Stabilizing selection functions to maintain the average type in a population. Disruptive selection is a procedure whereby both extremes of a population are selected. Beyond the different phenotypic responses to these three modes of selection, the genetic consequences also differ (Hartl

1980). Both directional and disruptive selection will lead to a diminution in genetic variability in the selected groups; with disruptive selection this will be in two different directions. On the other hand, stabilizing selection will tend to maintain the genetic variability present in a population. While each of these types of selection are possible, in natural situations the effect of selection will normally be to stabilize or normalize a well-integrated phenotype (Mayr 1970).

As can be seen from this rather brief consideration of some of the aspects of selection critical to genetic change in populations, it is a complex process. In addition, selection mechanisms work the same whether they are applied to affect genetic change through a conscious effort or through inadvertent or unintended processes. As a result of the exploitation and/or the augmentation of natural resources, selective pressures are often applied to populations that are neither the consequence of a directed human effort nor the result of totally natural influences. While it can be argued that these should be categorized as natural selection, the fact that some control can be exercised over inadvertent selection suggests that it should be considered separately and efforts made to understand and address the potential genetic consequences.

SELECTION IN DIFFERENT PHASES OF CULTIVATION

Precultivation phase

In this phase the primary concern is usually the maintenance of naturally reproducing populations and there is a tendency to disregard selection as a major factor in determining the genetic composition of these groups. However, two areas deserve some attention when considering populations that receive only minimal contribution from cultured fish.

First, natural selection is constantly acting on these groups whether the genetic effect is measurable or not. Numerous studies have been conducted to attempt to quantitatively assess the magnitude of natural selection and the associated genetic effects (Prout 1969, Christiansen and Frydenberg 1973, 1974, Christiansen et al. 1977, Buroker 1979). With few exceptions, these studies have yielded less than definitive results. The major problem encountered has been the estimation of the fitness of genotypes.

Natural selection can only be quantified by obtaining an estimate of genetic fitness, and the latter has a number of complexities that make it very difficult to estimate in natural circumstances (Hartl 1980). Consequently, while we can surmise that natural selection is operating within populations, it is very difficult to assess its magnitude.

The second area that needs to be considered in the precultivation phase of salmonid culture is the selective effect of the fisheries that harvest these populations. The selective impact of most fisheries is, in many ways, analogous to artificial selection imposed via controlled breeding of salmon and trout. Harvesting of fish imposes a strongly directional form of selection on salmonid populations because of the nature of the gear and management regulations. For example, Mathisen (1971) reported that 5¾-inch mesh gillnets caught five times as many sockeye salmon that matured after 3 years of ocean residence than after 2 years (both sexes combined) when these two groups were present in about equal numbers. Thus, the use of this size gillnet removes a disproportionate number of older (larger) maturing salmon from the reproducing population, effectively directing selection toward younger (smaller) maturing salmon. This, combined with the effectiveness of the modern commercial fishery, has the potential to exert a very strong selective influence on populations. Ricker (1972 and 1981) has argued strongly that the selective nature of the fishery causes decreases in size and age at maturity in several salmonid species in the Pacific Northwest. It is unlikely that the fishery exerts the consistent selection pressure that can be applied by a designed breeding program, but some of the changes induced by the fishery can have adverse effects on traits that are important to the productive capability of the stocks (Ricker et al. 1978).

The major uncertainty regarding the genetic effect of selective harvesting by the fishery is the definition of the genetic parameters that are altered. Changes that have been measured in fish can be explained by either (1) the selective elimination of populations; or (2) selection within a population of certain phenotypes (Ricker 1981, Alexandersdottir and Mathisen 1982, Nelson and Soule 1987). Both of these mechanisms result in genetic change, but only the latter is a continuing problem that can be addressed by changes in fisheries management. For example, management options to address the decline in size of salmon could include (1) reducing the allowable mesh size of gillnets; (2) curtailment of trolling; or (3) taking a substantial portion of the allowable catch near spawning grounds by gear that would counteract the effects of standard commercial gear (Ricker et al. 1978). Furthermore, recent evidence on salmonids indicates that variation within a population is the major reservoir of genetic variation for some species (Ryman 1983, Riddel 1986, personal communication). Thus, strong selection within a population could very effectively decrease an important source of genetic diversity, which is important to the population for adaptation to future environmental changes (Lewontin 1978).

The overriding problems in this phase of salmonid culture basically arise from a lack of definitive scientific data regarding the influence of natural selection, or selection due to harvest on salmonid populations. Consequently, in the precultivation phase information is needed on which to base management decisions. First, designed studies are needed to provide some objective assessment of the effects of various types of fishing gear and different regulations on salmonid populations. Most studies to date, although excellent in-depth analyses, have been observations of data with numerous variables that can only be marginally addressed. Second, efforts are needed to genetically define the traits that are impacted by selection within a population. Without such data, the effects of selection cannot be quantified. Finally, procedures for the regulation and management of the fishery need to incorporate a degree of flexibility in regard to characteristics that can have a strong directional selective effect on the impacted populations. The use of more variable gear types and regulations that diffuse the fishery in space and time should be considered.

Cultivation phase

This phase of salmonid culture differs from the previously considered phase, in some situations, only by the fact that culturists have added control over reproduction. Many of the same selection pressures still need to be considered, particularly where a population is being cultured for enhancement purposes. The potential for genetic changes due to selection can vary broadly depending on the procedures used to choose the adults contributing to the next generation. Selective effects are, in part, determined by the proportion of adults used

to contribute gametes to the next generation: the smaller and "better" the proportion selected, the greater the selection intensity and the greater the response will be in succeeding generations. This relationship is expressed by:

$$R = ih^2 \sigma_P^2$$

where R = Response (change in the population mean/generation)
i = Intensity of selection
h^2 = Heritability, and
σ_P^2 = Phenotypic standard deviation

There are a number of salmonid examples in the literature where biological changes have resulted from directed selection in this phase of cultivation (Embody and Hayford 1925, Davis 1935, Lewis 1944, Donaldson 1971). In most of these situations the changes were achieved by individual selection of fish exhibiting superior phenotypes, but with little documentation of the genetics of the traits selected or the genetic changes introduced. However, the efficacy of this type of approach for alteration of the phenotype in a directed manner is apparent, and recent work suggests that directed selection programs with salmonids will yield even further changes (Gjedrem 1983).

Another type of selection in this phase of cultivation may also yield major changes. The same characteristics that make salmonids responsive to selection in directed programs also make them susceptible to the pressures of inadvertent selection. When fish are placed into a culture situation, they encounter an entirely new array of circumstances (e.g., artificial food, high rearing densities, different temperature regimes, etc.) which amount to a series of selective pressures that can alter the genetic constitution of the population. While considerable effort has been invested in documenting phenotypic changes from hatchery rearing, no designed studies have been conducted to measure genetic changes that may be due to hatchery practices. Undoubtedly some genetic changes have occurred, but very little meaningful action can be taken to counteract undesirable effects until the genetics are understood.

Associated with these unknown parameters is a particular problem with populations that are reintroduced into the natural environment. The potential exists for the cultivated stocks and naturally reproducing stocks to be quite different. For the most part artificial selection, whether it is directed or inadvertent, is directional. This leads to a decrease in the genetic variability in populations, and genetic variability appears to be important for the continued viability of natural populations (Lewontin 1978). Also, artificial selection generally optimizes a limited set of traits, whereas in natural populations total fitness is the trait that is thought to be optimized (Kapuscinski and Lannan 1986) and fitness is a function of the entire phenotype. Thus, a multitude of traits are important to fitness. Consequently, interbreeding of the two groups will result in combining genetic material molded by very different selection pressures. The results from this occurrence are a matter of major debate and concern (Hynes et al. 1981, Krueger et al. 1981, Allendorf and Ryman 1987). Although other factors (e.g., genetic drift) can also influence the genetics of hatchery stocks, the selective results from culture practices and those from natural selection need to be defined and measured.

In summary, the cultivation phase of salmonid culture lends itself both to a directed selection approach and to inadvertent selection. Both have important implications when cultivated populations are mixed with natural populations. Immediate attention needs to be given to the quantification of the genetic effects of inadvertent selection occurring as a result of cultivation. Further, the genetic impact of these alterations on natural populations needs to be definitively measured. Many beneficial results from directed selective changes can be realized with salmonid populations in the cultivation phase, but much scientific data is needed for its proper utilization.

Postcultivation phase

Of the previously highlighted factors important to the genetic effects of selection in salmonid cultivation, about the only one not critical to this phase of culture is selection imposed by the fishery. Natural selection is still a factor, although it assumes a somewhat less important role in affecting genetic change. However, natural pressures can limit the magnitude of gains in a designed artificial selection program. In fact, natural selection has been implicated as the most likely explanation for a lack of response in some artificial selection programs (Pirchner 1969).

Selection will be a major factor in the success of a cultivation program where salmonids are raised under controlled environmental circumstances. For the most part, salmon and trout raised under these conditions will be destined for direct commercial harvest and, thus, subject to the constraints imposed for efficient aquaculture. As a consequence of these requirements, it is necessary to set defined and specific goals for improvement of traits important to production. In addition, the stocks will need to be adapted for specific purposes and it will be necessary to predict the gains that can be realized from genetic manipulation. All of these needs can be met via a designed breeding and selection program.

Because of the relatively recent development of programs with an emphasis on the selection of salmonids for commercial production, there have not been many opportunities to monitor and assess the progress of selection with these species. Probably the best documentation of results from continuing programs is found in work with common carp conducted in Israel and eastern Europe (Moav and Wohlfarth 1966, Bakos 1976). Selection work has been reported on specific traits in rainbow trout (e.g., Kincaid et al. 1977), Atlantic salmon (Gjedrem 1979, Gunnes and Gjedrem 1981) and coho salmon (Iwamoto et al. 1982, Saxton et al. 1984). However, few reports are available on long-term selection programs designed for large-scale production, with the exception of coho salmon (Hershberger and Iwamoto 1984).

Hershberger and Iwamoto (1984) devised a selection program with coho salmon to meet the needs of commercial production of pan-sized fish in marine net-pens. This included a definition of the traits of most importance to the economic production of fish in marine net-pens, determination of the genetic parameters defining these traits, and designing and testing a selection program that would maximize the genetic response of the fish and meet the requirements of production (Figure 2). This program has yielded genetic and phenotypic gains for the industry, and provided a broodstock exhibiting improved performance traits (Figure 3) in a semicontrolled rearing environment. Such genetic approaches will be necessary to the continued advancement of the commercial culture of salmonids.

There are several recognized problems in this phase that need to be addressed for selection to retain its usefulness. As mentioned earlier, selection can be equated with controlled or directed inbreeding. Thus, inherent to such a system is a

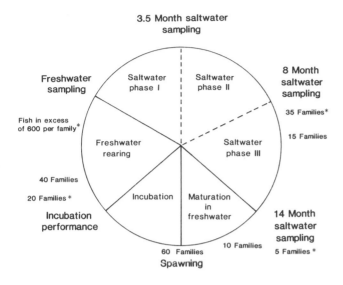

Figure 2. Selection scheme used for the development of stocks for the marine net-pen culture of "pan-sized" coho salmon.
Note: The complete circle represents the 2-year life cycle of these fish grown under an accelerated rearing regime, and the arrows indicate points of culling fish that do not meet selection criteria.

decrease in the genetic variability of the organisms involved. At some point this will lead to a diminished selection response and, perhaps, a decrease in productivity of the population. Consequently, selection programs must be designed with this eventuality in mind, and a plan to alleviate inbreeding problems. There have been numerous approaches proposed to address such difficulties (Kincaid 1983,

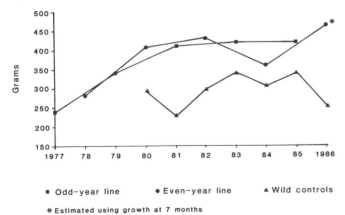

Figure 3. Average weight at 8 months of rearing in marine net-pens for odd- and even-year lines of coho salmon selected for five generations.
Note: Data for control lines of coho salmon from 1980 to 1986 are also presented for comparison with the selected lines.

Schom and Bailey 1986), but none of these has been tested in production circumstances. The only reservoir of "new" genetic variation available for potential use in these circumstances is that contained in natural populations. Thus, even with a strong emphasis on directed, artificial selection, there should be some effort to retain and protect natural stocks, both to protect and preserve the health of an important natural source, and as an extremely critical source of genetic variability for future directed selection programs.

SUMMARY

Selection, whether it is directed or inadvertent, can have major effects on the genetic composition of salmonid populations. The mechanism of action of selection is the same regardless of its source. However, the resultant product is very much a function of (1) the genetics of the traits involved; (2) the intensity with which selection is applied; and (3) whether there is any recognized or defined selection goal. Most of the problems within the various phases of salmonid production derive from a lack of adequate genetic information on traits of interest or perceived importance. This is the first problem to be solved. Second, the various sources of selection in different phases of culture need to be recognized and addressed on an appropriate level. The fishery, for example, could be the major selective agent in changing the genetic composition of some populations, and this can only be altered by management decisions. Finally, selection can be a powerful method to affect needed genetic changes in populations. This is most readily recognized in the culture of salmonids for commercial production purposes. However, other phases of culture may also be able to benefit from a defined, goal-oriented selection program. At the very least, "undesirable" genetic results could be eliminated by recognition of the major sources of selection, quantification of their effects, and taking the necessary steps to minimize their influence.

LITERATURE CITED

Alexandersdottir, M., and O.A. Mathisen. 1982. Changes in S.E. Alaska pink salmon (*Oncorhynchus gorbuscha*) populations, 1914-1960. Fisheries Research Inst. Report No. FRI-UW-8212. Univ. of Washington. 55 pp.

Allendorf, F.W., and N. Ryman. 1987. Genetic management of hatchery stocks. Pages 141-159 in N. Ryman and F. Utter, eds., *Population Genetics & Fishery Management*. University of Washington Press, Seattle.

Bakos, J. 1976. Crossbreeding Hungarian races of common carp to develop more productive hybrids. Pages 633-635 in T.V.R. Pillay and W.A. Dill, eds., *Advances in Aquaculture*. FAO Technical Conference on Aquaculture, Kyoto, Japan, 26 May-2 June 1976. Fishing News Books Ltd., Farnham, England.

Buroker, N.E. 1979. Overdominance of a muscle protein (Mp-1) locus in the Japanese oyster, *Crassostrea gigas* (Ostreidae). *J. Fish. Res. Board Can.* 36:1313-1318.

Christiansen, F.B., and O. Frydenberg. 1973. Selection component analysis of natural polymorphisms using population samples including mother-offspring combinations. *Theor. Pop. Bio.* 4:425-445.

Christiansen, F.B., and O. Frydenberg. 1974. Geographical patterns of four polymorphisms in *Zoarces viviparus* as evidence of selection. *Genetics* 77:765-770.

Christiansen, F.B., O. Frydenberg and V. Simonsen. 1977. Genetics of *Zoarces* populations. X. Selection component analysis of the Est III polymorphism using samples of successive cohorts. *Hereditas* 87:129-150.

Davis, H.S. 1935. Improvement of trout broodstock through selective breeding. *Prog. Fish-Cult.* 13:1-6.

Donaldson, L.R. 1971. Development of select strains of salmonid fishes for use in management. Pages 19-31 in W.F. Carter, ed., *Atlantic Salmon Workshop*. Special Publication Series, Vol. 2, No. 1, International Atlantic Salmon Foundation, New York.

Embody, G.C., and C.O. Hayford. 1925. The advantages of rearing brook trout fingerlings from selected breeders. *Trans. Am. Fish. Soc.* 55:135-148.

Gjedrem, T. 1979. Selection for growth rate and domestication in Atlantic salmon. *Z. Tierz. Zuchtungsbiol.* 96:56-59.

Gjedrem, T. 1983. Genetic variation in quantitative traits and selective breeding in fish and shellfish. *Aquaculture* 33:51-72.

Gunnes, K., and T. Gjedrem. 1981. A genetic analysis of body weight and length in rainbow trout reared in seawater for 18 months. *Aquaculture* 15:19-23.

Hartl, D.L. 1980. *Principles of Population Genetics*. Sinauer Associates, Inc., Sunderland, MA. 488 pp.

Hershberger, W.K., and R.N. Iwamoto. 1984. Systematic genetic selection and breeding in salmonid culture and enhancement programs. Pages 29-32 in C.J. Sindermann, ed., Proceedings of the eleventh U.S.-Japan meeting on aquaculture, salmon enhancement. Tokyo, Japan, October, 1982. NOAA Tech. Rep. NMFS 27.

Hynes, J.D., E.H. Brown Jr., J.H. Helle, N. Ryman and D.A. Webster. 1981. Guidelines for the culture of fish stocks for resource management. *Can. J. Fish. Aquat. Sci.* 38:1867-1876.

Iwamoto, R.N., A.M. Saxton and W.K. Hershberger. 1982. Genetic estimates for length and weight of coho salmon (*Oncorhynchus kisutch*) during freshwater rearing. *J. Heredity* 73:187-191.

Kapuscinski, A.R.D., and J.E. Lannan. 1986. A conceptual genetic fitness model for fisheries management. *Can. J. Fish. Aquat. Sci.* 43:1606-1616.

Kincaid, H.L. 1983. Inbreeding in fish populations used for aquaculture. *Aquaculture* 33:215-227.

Kincaid, H.L., W.R. Bridges and B. Von Limbach. 1977. Three generations of selection for growth rate in fall spawning rainbow trout. *Trans. Am. Fish. Soc.* 106:621-629.

Krueger, C.C., A.J. Gharrett, T.R. Dehring and F.W. Allendorf. 1981. Genetic aspects of fisheries rehabilitation programs. *Can. J. Fish. Aquat. Sci.* 38:1877-1881.

Lewis, R.C. 1944. Selective breeding of rainbow trout at Hot Creek hatchery. *CA Fish and Game* 30:95-97.

Lewontin, R.C. 1978. Adaptation. *Scientific American* 239:156-169.

Mathisen, O.A. 1971. Escapement levels and productivity of the Nushagak sockeye salmon run from 1908 to 1966. *Fish. Bull.* 69:747-763.

Mayr, E. 1970. *Populations, Species and Evolution*. Belknap Press of Harvard University Press, Cambridge, MA. 453 pp.

Moav, R., and G.W. Wohlfarth. 1966. *Genetic improvement of yield in carp*. FAO Fisheries Report Vol. 4, No. 44, pp. 12-29.

Nelson, K., and M. Soule. 1987. Genetical conservation of exploited fishes. Pages 345-368 in N. Ryman and F. Utter, eds., *Population Genetics & Fishery Management*. University of Washington Press, Seattle.

Pirchner, F. 1969. *Population Genetics in Animal Breeding*. W.H. Freeman and Co., San Francisco. 274 pp.

Prout, T. 1969. The estimation of fitness from population data. *Genetics* 63:949-967.

Ricker, W.E. 1972. Hereditary and environmental factors affecting certain salmonid populations. Pages 19-160 in R.C. Simon and P.A. Larkin, eds., *The Stock Concept in Pacific Salmon*. H.R. MacMillan Lectures in Fisheries, University of British Columbia Press, Vancouver, B.C.

Ricker, W.E. 1981. Changes in the average size and average age of Pacific salmon. *Can. J. Fish. Aquat. Sci.* 38:1636-1656.

Ricker, W.E., H.T. Bilton and K.V. Aro. 1978. Causes of the decrease in size of pink salmon *(Oncorhynchus gorbuscha)*. Dept. of Fisheries and the Environment, Canada. Fisheries & Marine Service Technical Report No. 820. 93 pp.

Ryman, N. 1983. Patterns of distribution of biochemical genetic variation in salmonids: differences between species. *Aquaculture* 33:1-21.

Saxton, A.M., W.K. Hershberger and R.N. Iwamoto. 1984. Smoltification in the net-pen culture of accelerated coho salmon *(Oncorhynchus kisutch)*; quantitative genetic analysis. *Trans. Am. Fish. Soc.* 113: 339-347.

Schom, C.B., and J.K. Bailey. 1986. Selective breeding and line crossing to reduce inbreeding. *Prog. Fish-Cult.* 48:57-60.

Wilkins, N.P. 1981. The rationale and relevance of genetics in aquaculture: an overview. *Aquaculture* 22:109-228.

Afterword

World harvest of salmon still comes mostly from naturally produced fish, but the contribution from aquaculture is substantial and continues to grow rapidly. There is a strong likelihood that aquaculture production will exceed natural production before the turn of the century. Japan is the recognized economic leader in ranching and Norway in farming.

Statistics for 1985 indicate that Japanese ranched salmon accounted for 18 percent by weight of the world harvest. Farmed salmon from all countries accounted for an additional 5 percent. The remaining 77 percent consisted of ranched salmon from countries other than Japan plus naturally produced salmon. Although there is no accurate way to separate contributions from these latter two sources, we can approximate their contributions. One approach is to compare smolt production from Japan with other countries and make simplifying assumptions about marine survival.

An unknown portion of the record 1985 harvest of 920,000 mt originated from approximately 4.4 billion smolts released from hatcheries. Two billion were from hatcheries in Japan and 2.4 billion from hatcheries elsewhere. Smolt-to-adult survival of Japanese ranched salmon has increased to about 2.5 percent in recent years. Assuming that averge survival is about the same elsewhere, Japan accounted for 50 million ranched adults and other countries for 60 million.

Mostly chum salmon are ranched in Japan, whereas about 40 percent of salmon ranched by other countries are smaller pink salmon. Adjusting for adult size, my estimated percentages of total biomass from aquaculture and natural production in 1985 are:

Aquaculture	
Farmed	5 percent
Ranched in Japan	18 percent
Ranched elsewhere	16 percent
Total	39 percent
Natural	61 percent

Recent exponential growth of salmon aquaculture has resulted in production of farmed salmon doubling in about 3 years and of ranched salmon in about 10 years. Economic and/or biological limitations are likely someday to mitigate against a continuation of these high rates of growth, but I expect world growth to continue at present rates for another decade at least.

Competition for existing markets is becoming more intense. But there also is increasing awareness among consumers that salmon offer superior health benefits when contrasted with many other types of animal protein. There may be great potential in the market for consumers to substitute salmon for pork, beef, and possibly other sources of animal protein.

Costs for producing a ton of ranched or farmed salmon are likely to decline as production increases, and the price of salmon could become highly competitive with other sources of animal protein. Reduced costs are anticipated from economies of scale, technological innovations, and increased survival rates. Competitive market pressures are expected to stimulate cost reductions and development of new products. Processing and marketing are likely to become more centralized.

Allocation of ranched salmon among user groups will continue to be an important policy issue. The salmon rancher makes use of a common property resource (the ocean) to pasture fish. Harvest of a reasonable portion of ranched salmon by a public fishery is one means to insure that the rancher pays economic rent for use of the ocean. Sport fisheries are especially attractive for this purpose because the opportunity for sport fishing can be made available to all citizens at reasonable cost.

Fishery managers will continue to face problems on overharvest of wild salmon where fisheries harvest hatchery and natural stocks concurrently. Hatchery stocks typically tolerate a much higher harvest rate than natural stocks because far fewer spawners are required to produce a given number of smolts in a hatchery than in nature. However, it is possible to reduce harvest of wild fish while increasing the total harvest of hatchery plus wild fish in mixed-stock fisheries. To accomplish this, managers need to hold the harvest rate at a fixed percentage of the total supply while increasing the production of hatchery fish. The dilution of wild with increased numbers of hatchery fish will insure a reduced harvest rate on wild fish while the total catch increases. Under this scenario hatcheries will receive large surpluses of adults which potentially can be sold to pay production costs.

It is well documented that a significant portion of hatchery salmon stray into natural spawning areas. Interbreeding of stray hatchery fish with wild fish raises biological questions that are important to conservation of natural stocks. Interbreeding provides new genetic material for a population. This may be beneficial or detrimental, depending on circumstances. Effects on fitness for survival of resulting progeny are usually unknown or are inferred from theory or assumptions. Techniques capable of reducing incidence or genetic effects of interbreeding need to be developed and/or evaluated. Some recognized approaches include:

— Develop hatchery stocks which mature earlier or later than wild stocks.
— Use sterile hybrids and/or polyploids for ranching.
— Use imprinting techniques which reduce straying.

These and possibly other approaches to reduce interbreeding or genetic effects of interbreeding merit priority consideration for research and development.

Capacity of marine waters to grow salmon is receiving increased attention from the scientific community. The most intensively ranched coastal waters extend from about 40° to 50° north latitude in the northwestern Pacific. The area in question encompasses northern Honshu, Hokkaido, and southern Sakhalin islands. At least three billion juvenile salmon are released annually. Survival of Japanese chum salmon, which constitute two-thirds of the total number released, has shown steady improvement in spite of increased numbers of smolts.

Survival of hatchery and wild stocks has been highly variable in the northeastern Pacific. Survival of ranched coho declined precipitously off California, Oregon, and Washington in the late 1970s and early 1980s. Hatchery releases of coho smolts increased from about 40 to 60 million in this period, but environmental variables unrelated to density of smolts have been recognized as potential causes of reduced survival. In Bristol Bay, Alaska, on the other hand, smolt-to-adult survival of naturally produced fish increased spectacularly in the late 1970s and early 1980s. Numbers of adults returning to Bristol Bay have averaged 44 million in recent years. Coastal waters of Bristol Bay encompass only about one degree of latitude, and smolt numbers have ranged upwards of 900 million. The Bristol Bay experience, along with those of Japan and the U.S.S.R., suggests that marine waters have the capacity to grow a much larger biomass of salmon than currently realized. Nevertheless, limitations to the capacity of marine waters to grow salmon are likely to become evident sometime in the future if trends for increased release of hatchery smolts continue.

Improved technologies for salmon farming and ranching are contributing to reduced costs and increased survival. The potential for these technologies is substantial. It has already been mentioned that smolt-to-adult survival of ranched salmon averages 2.5 percent in Japan. The corollary is that marine mortality is 97.5 percent. Thus, a 1 percent decrease in mortality from improved fish husbandry, release strategy, or some other aquaculture procedure translates into a 40 percent increase in survival and production, possibly with only a modest increase in cost.

Future structure of salmon aquaculture will be molded by politics as well as by biological and economic factors. Some sovereignties largely allow market forces to dictate growth of farming or ranching or both. Others erect political barriers to growth, including prohibitions against participation by nongovernmental institutions. Policies affecting salmon aquaculture are typically transient, and there has been a general, but not universal, trend toward liberalization to encourage economic growth and diversification.

Policy decisions affecting salmon aquaculture are based largely on biological impacts, economic trade-offs, and political expediency. Most of the biological, economic, and policy issues raised and discussed in these proceedings of the World Salmonid Conference will receive continuing attention from those who grow, harvest, market, and manage salmon and those who establish policy affecting their production.

William J. McNeil

Appendix

Common and Scientific Names of Fish Cited

Alewife	*Alosa pseudoharengus*
American eel	*Anguilla rostrata*
Atlantic salmon	*Salmo salar*
Belukha whale	*Delphinapterus leucas*
Black rockfish	*Sebastes melanops*
Blue pike	*Stizostedion v. glaucum*
Brook trout	*Salvelinus fontinalis*
Brown trout	*Salmo trutta*
Carp	*Cyprinus carpio*
Channel catfish	*Ictaluris punctatus*
Cherry salmon	*Oncorhnychus masou*
Chinook salmon	*Oncorhynchus tshawytscha*
Chub (bloater)	*Coregonus hoyi*
Chum salmon	*Oncorhynchus keta*
Coho salmon	*Oncorhynchus kisutch*
Cutthroat trout	*Salmo clarki*
Deepwater sculpin	*Myoxocephalus quadricornis*
Emerald shiner	*Notropis atherinoides*
Halibut	*Hippoglossus stenolepis*
Herring	*Clupea pallasii*
Kamchatkan trout	*Salmo mykiss*
Kokanee (see Sockeye salmon)	
Lake herring	*Coregonus artedii*
Lake sturgeon	*Acipensev fulvescens*
Lake trout	*Cristivomer namaycush*
Pacific salmon	*Oncorhynchus sp.*
Pacific sand lance	*Ammodytes hexapterus*
Pacific whiting	*Merluccius productus*
Pink salmon	*Oncorhynchus gorbuscha*
Rainbow trout (see Steelhead)	
Sea lamprey	*Petromyzon marinus*
Slimy sculpin	*Cottus cognatus*
Smelt	*Osmerus mordax*
Sockeye salmon	*Oncorhynchus nerka*
Steelhead	*Salmo gairdneri*
Walleye	*Stizostedion vitreum*
Whitefish	*Coregonus clupeaformis*
White perch	*Morone americana*
Yellow perch	*Perca flavescens*

Index